Sound of Black Switzerland

Jessie Cox

Sounds of Black Switzerland

Blackness, Music, and Unthought Voices

DUKE UNIVERSITY PRESS
Durham and London   2025

© 2025 DUKE UNIVERSITY PRESS. All rights reserved
Printed in the United States of America on acid-free paper ∞
Project Editor: Michael Trudeau
Designed by Courtney Leigh Richardson
Typeset in Minion Pro and Helvetica by
Westchester Publishing Services

Library of Congress Cataloging-in-Publication Data
Names: Cox, Jessie, [date] author.
Title: Sounds of Black Switzerland : blackness, music, and unthought voices / Jessie Cox.
Description: Durham : Duke University Press, 2025. |
Includes bibliographical references and index.
Identifiers: LCCN 2024023858 (print)
LCCN 2024023859 (ebook)
ISBN 9781478031437 (paperback)
ISBN 9781478028215 (hardcover)
ISBN 9781478060420 (ebook)
Subjects: LCSH: Uzor, Charles, 1961—Criticism and interpretation. | Black people—Switzerland—Social conditions. | Black people—Switzerland—Music. | Racism against Black people—Switzerland. | Music and race—Switzerland. | Music—Political aspects—Switzerland. | Music—Social aspects—Switzerland.
Classification: LCC DQ49.A49 C68 2025 (print) | LCC DQ49.A49 (ebook) | DDC 305.896/0494—DC23/ENG/20241118
LC record available at https://lccn.loc.gov/2024023858
LC ebook record available at https://lccn.loc.gov/2024023859

Cover art: *White Gaze II, Black Square—Mirror IV*, small iteration, 2024. © Akosua Viktoria Adu-Sanyah. Courtesy of the artist.

Contents

Acknowledgments — vii

Introduction.
Black Swiss — 1

1. Interstitial Listenings I:
Charles Uzor's *Bodycam Exhibit 3*, Part I — 17

2. Blackness and Black Lives in Switzerland — 33

3. Interstitial Listenings II:
Blurring the Hold — 59

4. Afrofuturist Archeology:
Citizenship and the Delimitation of Life with Death — 71

5. Interstitial Listenings III:
Black Music behind the Wormhole — 91

6. Mothership Connections — 101

7. Interstitial Listenings IV:
8'46" George Floyd in Memoriam and *White Gaze II Black Square* — 121

8. Listening with Black Switzerland   137

9. Interstitial Listenings V: 159
Charles Uzor's *Bodycam Exhibit 3*, Part II

10. Black Life / Schwarz-Sein   167

Conclusion.   187
Alongside a Chorus of Voices

Postface.   197
Endless Endlessness

Notes   203

Bibliography   225

Index   243

Acknowledgments

I extend my deepest gratitude to all those involved in the music that made this book, from the creators, performers, composers, listeners, curators, activists, staff, administrators, and many more. This work benefited immensely and is indebted to conversations with Charles Uzor about his life and our music. I thank those who have opened the possibility of these words, these thoughts, by way of continued conversation, particularly during my time at Columbia University: George E. Lewis, Ellie Hisama, and participants of the seminars they led, as well as Sam Yulsman, Isaac Jean-François, and Hannah Kendall with whom I have directly conversed around these and related topics.

The text has greatly benefited from the scholarly communities with whom I have been in conversation over the course of writing, including the Music Department at Columbia University; the Comparing Domains of Improvisation speaker series at Columbia University in 2022; the Conference of the Forum of the International Association for Word and Music Studies on Words, Music, and Environment at Chatham University, Pittsburgh, also in 2022; the annual meeting of the American Comparative Literature Association at the National Taiwan Normal University in 2023; and the AMS/SEM/SMT joint annual meeting in New Orleans, also in 2023.

A very early draft of chapter 2 received feedback from Robert Gooding-Williams, and I am thankful for his invaluable comments. I would like to thank the peer reviewers who provided invaluable feedback to this manuscript's earlier drafts. A special thanks to Ken Wissoker and the team at Duke University Press for bringing these sounds on pages to your ears.

Lastly, my deepest gratitude and admiration to my partner, Lucy Clifford, for her unwavering support in my endeavors, and to my family: Marianne, Roy, Jemina, Hedi, Julie, Mick, and the extended family. And, of course, there are so many more, so many I can't name all, but who were and are my continual accompanists; while writing, composing, making music, whether in sound or as text, they are those who make the music possible.

Introduction
Black Swiss

Many Beginnings

Changing our ears means listening beyond what can be heard. This book is about that kind of musical practice. It is clear to me now that writing this book, this work of placing ink on the page or fingers on the keyboard, with its rattling and percussive sound, is not only an extension of my musical practice but part of it. This might seem strange in some ways. How could the rather dull sound of these buttons be related to musicians playing instruments? How can staring at a mute piece of paper be anything like the flourishing sounds of musicians in concert? It is through an engagement of what musicians do, what they do to themselves and the world, by teaching us to listen, that music commences to ring in our ears, to listen again and again, like the enduring and never-ending rehearsal of our sound. That to me is the lesson of music. It is in this sense, in this ear-transforming sounding that I find also the core concern of this book. What is this sound? What are we learning to listen to through these notes before our eyes? Sounds of Black Switzerland.

This book is a *case*—an offering as well as a petition—of and for Black Swiss lives, to think about Black lives in Switzerland. I want to think about how we make audible, how we bring to thought, that which has been (placed) in silence. This sound is coming to us "in a silent way," to reference jazz icon Miles Davis, but is louder than loud.[1] Its appearance seems unthought, but it has been here all along. To think and write about Black Switzerland was for me a kind of first; it was a kind of new thought that I had to find myself. While it is clear that Black Swiss (and this term, or name, itself) have not been thought about much—a fact that seems at the time of writing this book to be changing—it is still somewhat difficult to relay this experience of having to name something that has claimed, or is claiming, not to exist. When you are a Black Swiss, a common experience is that you are told you are either Black or Swiss.[2] I knew the term *Black* and I knew the term *Swiss*. But Black is commonly thought of as outside of Switzerland, and, as a result, Blackness is taken from Black lives in Switzerland to make them Swiss. And/or Black Swiss are marked as foreigners within where they are from or where they have citizenship or live (i.e., Switzerland).[3] The lived experiences of Black lives in Switzerland are erased. Black Swiss lives should not exist.[4] But I am living as a Swiss and I am Black; Black and *living* in Switzerland—*Black Swiss*. Thus, I began this book before I was able to think it, before I was able to think *Black* and *Swiss* in conjunction in excess of a mixture, and it is this sound-making, as the writing on paper, that brings these sounds to an ear that must keep learning to listen. Here I want to focus on this practice, the practice that Blackness demands: to keep listening without claiming to know it all. For if we can change our ears, if we can learn to hear lives in excess of a fabricated proper life here (a proper citizen from here), then maybe we can imagine another world, one where Black lives matter.

In some ways, this book is about beginnings, but beginnings become a kind of uncovering, where there are multiple beginnings, because in these contexts, where our sound is impossible, we must keep insisting on how there is sound in silence. This means it is not only about revealing antiblack structures but also about how lives are shared and experienced in excess of being denied a voice. While there are a number of groups and initiatives, many of which have received further push from the globality of the Black Lives Matter protests in 2020, there is still much work to be done.[5] Despite the existence of Black Europeans since before modernity, the notion of a white Europe has been manufactured, and not only socially or symbolically, but also materially enforced through various methods.[6] And even if there are more and more now, who hear this sound more readily—the sound of Black lives, of Black

Europeans, of Black Swiss—I want to elaborate how the experience of having no knowledge of it, of having no way to think my own lived experiences, not only teaches us something about antiblackness but also Blackness and the world. I may speak of it as another way of listening.

I know there are many like me, Black Swiss, or people who care about Black lives, also in Switzerland, who have (or had) no language, no way of speaking or thinking about the situation of Black Swiss. While discourses around antiblackness seem to grow in the aftermath of the national and international uncovering of police brutalities, I want us to think in excess of racism and antiblackness; I want to think about Black lives and the stories they tell, about what kind of radical potential to refigure the world lies in the study of their thought and practices, in this black study. If antiblackness denies our talking about Black lives, then to share our life experiences with each other in excess of what is supposedly properly here is black study's protest-celebration. The study of lives, stories that should not exist, is what I want to engage in, and this study is about us and the world. It is about asking what "us" sounds like, because antiblackness cuts off what we are, what Swissness looks like, what Blackness can't be. So, on the one hand, antiblackness demarcates and delimits what a proper citizen, what life itself, looks and sounds like. On the other hand, antiblackness also encircles blackness. Antiblackness denies Black lives to meet each other in excess of controlled enclosures. Blackness opens the question, the question of who "us" is, without the possibility of ever closing it. It bespeaks relations beyond what can be accounted for.

In 2020, Black Swiss artists wrote a public letter to all art institutions in Switzerland, asking them "to invest more actively in the necessary process to become more anti-racist."[7] They make it quite clear: antiblack brutalities claim lives also here in Switzerland:

> Within the last years, at least three Black men have been killed by the police in Lausanne and Bex: Mike Ben Peter, Lamine Fatty and Hervé Mandundu. None of their murderers have been convicted and therefore no justice has been awarded to these men or their families. One must also bring to light the fact that many assaults due to racial profiling by the police rarely end with the police being charged. The most prominent cases are those of Mohamed Wa Baile and Wilson A.
>
> While these examples speak to some of the most extreme forms of racism, we must acknowledge that anti-Black racism is a direct derivative of white supremacy: an oppressive system of beliefs and

discriminatory set of biases that is inherent in all structures of the Western World.[8]

Their call to action to art spaces is amplified by a recent call to the whole of Switzerland by a team of experts from the United Nations (UN), who have declared that the issue of antiblack racism in Switzerland is "urgent."[9] "Switzerland Refutes the Accusations of Racism by the UN," reads a headline in the national news, following the publication of this report. The basis for such refusal plays into the very mechanisms of antiblackness: that it bespeaks only individuals, and that the report was based on many "assumptions."[10] It is the turning into an exception of events of violence against Black lives that denotes their deaths as accidental and in turn also as outside of concern for study. In other words, the denial of antiblackness reifies and articulates the claim that these deaths are insignificant for study, for an investigation into the reasons behind why they happen, with such also inhibiting the prevention of further deaths and the asking of questions about the causes for them (see chapter 2).[11] The world keeps reproducing itself as antiblack in the very evasion of the question about Black lives and the violence inflicted on them.[12]

One of the artists who signed this open letter, Soraya Lutangu, who goes by the stage name "Bonaventure," created an EP thematizing this violence. It was after her nephew's death that she began working on this EP, and that made her engage antiblack racism. As she states: "It sounds so stupid now, but before that [the death of my nephew] I was fully focused on graphic and book design and I was obsessed with my career. Each time I would have those racist micro-aggressions (that are really common where I grew up) I would laugh about it. It was a way to protect myself, but today I'm not the same person at all. I am convinced that my nephew died also because of the colour of his skin, and this really shocked me, changed me."[13]

The opening track "Supremacy" tackles the problem head-on.[14] It reveals a violence usually willfully disassociated with Switzerland. The static sonic tapestry in combination with crunching, screeching, and pulsing sounds reflects a tenseness, a state of terror, that antiblack brutalities bring with them. But it is also a protest. Bonaventure layers selections of an interview by rap artist and activist Sister Souljah on top of this *state-of-emergency*-like soundscape. In it comes to the fore, as protest, how *Black* is a site for reclaiming not only one's own life but also the world, history, and study. How underneath, in spite of, against, and in excess of antiblackness there are Black lives, Black thought, and Black stories.

It is in this that comes to the fore how blackness bespeaks something that exceeds the confines of a world rooted in antiblackness. Blackness is the very possibility of thinking beyond the already given context of a world that both claims itself against otherness and attempts to keep erasing this antagonistic relation. What if "black" is the heading under which we can hear an (unavoidable) impossibility—the changing of the world towards the unknown, the unimagined, the speculative? By this I also mean that thinking about Black lives is like pushing the "anti" of antiblackness out, like turning the problem into study.[15] It is how by revealing the violence that enforces a normative society against those who must embody danger (i.e., Black lives), there comes to the fore lives lived in excess of brutal control over who *us* may be, who we may make relations with, who may live here, or in general. It is through this music that we learn of Black lives and their stories, where we engage in black study. Through our listening to this EP, we learn about antiblackness experienced in Switzerland, in the world, but also about Black lives. Bonaventure speaks in sound, and through this music our listening may exceed discourses that are entangled in the erasure of Black lives.

Switzerland as a country is connected to different spheres of articulations of Black studies due to its four language regions. The study of Black lives in Germany took its incipit in the '80s with Katherine Oguntoye, May Opitz, and Dagmar Schultz and their seminal book *Showing Our Colors*. The book is a space of meetings among lives, Black lives, and lives close to Black lives—it was the study of Black lives that caused also their meeting. In fact, the very term *Black* cannot exist without Black lives meeting, meetings that at once move past borders of an antiblack world and at the same time open up the possibilities of who *us* may be. Antiblackness controls and limits who may meet, who may make relations, and with whom one may associate. Blackness thus bespeaks making new relations—in life, in thought, in sound, and so forth.[16]

This also means that Black studies, and these terms, must exceed their instantiations, their narratives that emerge in antiblack structures. This is to say that the way in which *Black* becomes reduced to a nation such as the United States, or the reducing of Black lives to certain geographic locations, serves to delimit and erase Black *lives*.[17] These terms like *Black*, or *Afro-German*, must be in excess of these conditions that enclose blackness into distinct categories pre-given by an antiblack world. *Black* bespeaks the dealing with the contexts of the world as a way to change the world. For it is through these terms (*Black*, *Negro*, *Afro-German*, and *Black Swiss*) that antiblackness may be addressed, that the histories of those who fought against antiblackness may be told, and that the world may be transformed for the better.[18]

Black study in Switzerland has been practiced in the underground and in plain sight. Cikuru Batumike's three books, *Présence africaine en Suisse* (1993), *Être Noir africain en Suisse* (2006), and *Noirs de Suisse* (2014), mark a first in the thematization of Black lives in Switzerland. His initial two books focus on the immigration experience of Africans but his third monograph begins centering the particular situation of Black Swiss.[19] Batumike is a journalist, essayist, and poet who left Bukavu in then Zaire, and today the Democratic Republic of the Congo, in 1982 after having been imprisoned due to his political criticism, and has lived in Switzerland since 1984.[20] Thus, while the study of Black Switzerland has not received much attention, there have nonetheless been incursions by individuals against all odds.[21]

Slowly there is at least more focus on antiblack racism, especially in Switzerland. Exemplary is the very recent edited collection *Un/Doing Racism*, published in 2022, which thematizes a variety of modalities of racialization and unpacks Switzerland's relation to race and racism. While we do have these works, and many more, and they keep coming, the situation in Switzerland requires us to keep learning to learn to listen (because the sound keeps changing) to study Black lives. Similarly, the authors of *I Will Be Different Every Time*, published in 2020 to the library of black study in Switzerland, demonstrate that it is about the stories of us, of Black lives, of Black lives in Switzerland, and about how our stories make us; they are our sound.[22] Black study has been practiced in spaces and discourses beyond what they could know or bespeak, in music and art, where we may listen for some radical thought-work.

I begin this book with an elaboration of cases of antiblackness in Switzerland (chapter 2), how the specific case of Switzerland uncovers antiblackness's fundamental modalities of color-blindness and multiculturalism, followed by a more general thinking through of antiblackness as it relates to the formation of the nation-state (chapter 4). I do so to facilitate a *dive* into Black Switzerland and its sounds.[23] It is only through engaging the study of Black lives, and of blackness in general, that antiblackness can properly be critiqued. If antiblackness is the condition of erasing blackness from within (let's say, Switzerland), then any critical engagement that stops at the critique of antiblackness is still held in a bind with an antiblack world. As Switzerland teaches us, antiblackness polices the inside of the nation, not only in terms of physical geographic borders but also in terms of biological, *cultural*, and otherwise. This policing is an attempt to erase blackness from within. But in critiquing scenes of antiblackness come also to the fore those lives, those voices, who speak against antiblackness, who live despite antiblackness, who thrive with and in Blackness.

What I want to talk about is Black Swiss lives. I want to talk about the sounds coming from Black Switzerland. Writing these words on pages, I hear how my ears change, how the idea of Black Swissness does something to these enclosed categories. These names turned into something final and fixed. We must stay with this moment, this interstitial space, this temporal break, where the boundaries of enclosed spaces become audible. This is what it means to think and speak about blackness.

Speaking of Black Swiss

How do we speak about Black lives in Switzerland in excess of discourses, and terms, that are entangled in antiblackness's erasure of Black Swiss? This does not mean that there aren't people who have termed, or named, Black Swiss, who have formed organizations to think and speak about Black lives in Switzerland. Rather, I need to unravel a general situation of having to re-invent names because antiblackness makes *Black Swiss* into something that does not exist. Naming Blackness is about learning: about the world, oneself in relation to others, and coming together with others, being together with the world and others. In Blackness coming together, expressed as the out-of-bound naming of Blackness, like naming Black Swiss, as a way to speak about Black lives, and think about blackness, is a breakdown of names too. This is because such naming does violence to an economy of names based in proper relations—that is, names authorized by a world defined against black. It is a question of naming Blackness because: What do we do with something that has many names and no name at the same time? How do we even think about this lived experience, this fact, that is Black life?[24] For us to come together as a study of Black lives is to make Black lives matter, and such is the sound of lives in excess of what can be accounted for. Black, we must conclude, is not reducible to a name, even though we enter it into the world of names and words, but rather it is an excess of naming that unnames and names anew, where names aren't in themselves. Our situation is one without words for us, so we must make them ourselves. This is the musical creativity of Blackness. Blackness is stealing away from its name, as the continual changing of our sound. How do we come to make a sound? First, we must listen—listen to us, to each other, to the world—beyond what we thought sound could be.

Searching for a sound means thinking about music: whenever I compose music, I search for sounds in many ways, and I make sounds in this search that is the act of writing down notes on pages. Listen beyond what can be seen. Listen to the inaudible sounds beyond what could ever be heard.[25] It is like when you enter the concert hall and the orchestra begins to tune, but this time, instead of tuning to a prefixed pitch, like A, musicians, and us, are listening to each other to hear what we might have ignored in us.[26] When composing, I don't write what was already there; rather, I try to listen for what has escaped my ears, changing, in the process, how I listen. It is about re-learning to listen, again and again, and again. Like when we first learn a word, when a word is an ambiguous sound that we can't quite hear yet, and its meaning is blurry to our mind, although it already bespeaks a complex set of relations (in excess of already existing relations). Can we keep coming back to this moment, between spaces of knowing, where we really listen and change our own listening in the process?

Listening to the sound of Black lives in Switzerland began for me, of course, with finding my own sound—that is the process of becoming a musician. It was the moments in which I met (with my ears) Black Swiss and their sounds that thoughts of Black Switzerland came to me. But this change is illusory—this idea that I was ever listening, or sounding, on my own. For the sound that is me, the finding of my sound, is the process of hearing the meeting of others as me. I hear this me as a coming together of sounds, like music, that opens thought. I hear this book as an extension of this meeting, and as such, it is also opening a space to think about Black lives. It is in this listening beyond what is already accounted for, what I call *interstitial listening*, that our ears are transformed into instruments to join the music. In this book, there are five "Interstitial Listenings" chapters. These are the odd-numbered chapters (1, 3, 5, 7, 9). In them, we listen to the theorizing and thinking that happens in the music of Black Swiss. These are interspersed between the even-numbered chapters (2, 4, 6, 8, 10), which contain theoretical elaborations that are derived from and extend these listenings. We meet Black Swiss in listening to their music, as a relearning to listen, as an interstitial listening.

My use of the term *interstitial* is not a simple in-between space that can be reduced to its proper position as between wholes. The interstitial in this book is both a theoretical device and a compositional method for opening us to improvising new music. Throughout the book I shift our listening, like in a musical composition/performance.[27] The silence, or nonexistence, I show is not the absence of sound or the negative of existence. Rather, silences make sounds. Silence is not the end, or the death of things as the losing of a life that can be owned. Rather, it is death, the unthought, that which could be, what could have been, and all the possibilities never heard of. Silences are the dreams that make us; silences are sounding music. To hear Black Swiss lives is to encounter the unknown in a fashion that makes what is known unknown, which breaks down the boundary that in-between is supposed to uphold. That is because the term *Black Swiss* does not describe a self-contained, nor existing, or connected, officially recognized community, and so we must think beyond a world that denies Blackness's existence beyond being delimited by antiblackness.[28]

I center a series of musical works bespeaking Black lives in and as Switzerland. While there are numerous Black artists, musicians, thinkers, organizers, and activists in Switzerland, I decided, for this project, not to make a list. This is not because I do not believe in list-making (and I do hope others will continue uncovering Black Swiss musicians). On the contrary, I think that the very act of making lists that are written for and by Black lives, to document them and their lives or actions, can be radically disruptive activism (see chapter 3). This is because antiblackness requires Black lives to be erased, to disappear from here, to have no part, to be unaccountable, except as something that can be owned, or held, in contradistinction, so as to establish proper belonging and authority, through owning others. While there have been lists, lists of the value of Black slaves for example, that have been used to instantiate antiblack structures, these lists turn Black lives into only those kinds of numbers that track capital or serve the narrative that Black lives are no lives at all—an attempt to encircle blackness and hold it off. This bespeaks less a kind of fundamental flaw in numbers, lists, or other kinds of measurements, but rather a specific modality of limiting what can be registered. I will unpack this further in chapter 8 as also a shift in how we listen, and it will become evident how a different modality of thinking about registering is required to talk and think about, and with, Black lives. This also means that the absence of lists is far from a guaranteed counter to antiblackness, as the situation in Switzerland unequivocally proves.

Nonetheless, my approach here, in this work, is more toward trying to follow a theoretical thrust in a couple of artistic works; that is, this refiguring of what it means to count—we might say, of how to tell our stories. To this aim, I have included, firstly, a series of works that think through the situation of Black lives in Switzerland. Secondly, these artists' and their works' relation to Switzerland is not based on a kind of delimitation of belonging that is entangled in antiblackness. In other words, belonging is not reducible to being born somewhere, to naturalization, or other kinds of proper claims to citizenship that are based in the exclusion of those lives deemed improper. Black Swiss live here, migrated here, were born here, and all other ways of arriving/departing we might have. In this sense, it is about rethinking what it means to be a citizen, what it means to belong without being a citizen. We might say that this is a question of how we engage space and the world, as a question of the relations between each other, or us, and how such a question is not reducible to a proper belonging measured by a system of legality and rights to be able to be in a space, or being part of a space. Thus, for me, thinking about *Black Swiss* has never been about a kind of nationalism that emerged in the nineteenth century, a nationalism entangled in the protection of capital and wealth accumulated through exploitation of Blacks and Indigenous in particular. Rather, I hope to open the door with this term—which is really inevitable, as this term *Black Swiss* cannot not refigure the very notion of citizenship and belonging—a question of how we relate to each other, space, time, planet earth, and the cosmos.

While global Black Lives Matter protests reinvigorated discussions about antiblackness all over the world, they also allowed long-brewing initiatives to come to the fore and gain new steam. These were pushed forward by people working behind the scenes toward a more just world for many years. One of those coincided by mere chance with 2020—namely, the Afro-Modernism in Contemporary Music conference presented by Ensemble Modern, who are based in Frankfurt, Germany, in collaboration with my then doctoral advisor, Columbia University professor, composer, scholar, and pioneer of electronic music George E. Lewis. This conference, of which I was also a part, opened doors in the classical music world, particularly the one that cares about living composers and artists (as opposed to focusing mainly on a death canon). With such, it, and the events or changes following it, also opened my own ears. It was there that I began needing to use the term *Afro-Swiss*, which I turned into Black Swiss, so as to speak about my own experience.

But the first time I had used the term was when I had to reflect on the musical collaborations between me and Jérémie Jolo (whose work I speak about in chapter 3) for a presentation at Columbia University. I had to verbalize how our music bespeaks our lived experiences—to be Black and Swiss, in excess of given ways to think about Black lives or Swissness (or Europeanness). It also reflects our particular experience of antiblackness; for example, of how the divide between classical music and jazz (and pop, rock, and so forth) is entangled in questions of antiblackness and how we navigate these spaces.[29] In some ways, we can say that the music came first, before our recognition of a term to be able to reflect on it *extra-musically*. Music was our thinking in excess of readily available terms, discourses, and spaces. We were theorizing and studying blackness, Black lives, outside of possible discourses and thoughts. This is how I arrived at the idea that there might be a way to elaborate on the complexities of Black Swiss lives through studying their/our music.

Something popped in my ears at these musical meetings, like when your ears get used to the pressure change when diving into the waters or taking off in a flying object across different mediums. But my ears aren't alone; they listen with the world, and so it is things that happen in the world, such as the Black Lives Matter protests, that change how we listen. Shortly after 2020 I was introduced to the music of a Swiss composer who'd been composing radical sounds for years in St. Gallen, Switzerland. Charles Uzor is a Swiss Nigerian composer, and the discussion of two of his works build the cornerstones of this book (chapters 1, 7, and 9). Both of these works are homages to George Floyd, and both of these works not only critique antiblackness in general but also the form in which it tends to present itself in the classical music world and in Switzerland. At the same time, these two pieces also move beyond critiquing antiblackness to a celebration of Black lives and blackness's radical potentialities for the world and lives. These pieces, and Uzor's collection of works in general, brought me the idea of writing this book about Black Switzerland, because of the depth of the theorizing that I found within them. What Uzor's works for Floyd evince is that Black lives require thinking and living with the global as well as the provincial, the individual and the societal, the universal and the particular. Blackness is about the particular, the local places, individual people, without ever not also being a cosmic question.

While Uzor marks the incipit for this book, Swiss-based German Ghanian artist Akosua Viktoria Adu-Sanyah's artwork *White Gaze II Black Square* is where I found its conclusion. But this conclusion is not a closing of the book or the questions and research directions it opens. Rather, it is a kind of opening, like how a black hole's center, its end, is where it really opens time and space to the unknown (I discuss her piece in chapter 7). It marks the particular situation of Black Switzerland, or Black Europe, as a general question of blackness and the world, and particularly our current historical time.

"White gaze" is a term that comes from the seminal work of Martiniquan philosopher and psychologist Frantz Fanon for describing the antiblack gazing that he experienced when arriving in France.[30] While I unpack the question of this gaze further throughout the book, I want to highlight here simply how this gaze is about fixing negative judgment in blackness and, in turn, this blackness in some people. But Adu-Sanyah is uncovering, with her doubling down on the gaze, something about gazing, about looking (which is that glossiness, or a kind of excess of reflection and absorption, of blackness), where looking becomes the falling over and into the artwork of the observer.[31] This black square that is so glossy that it is full of noise—it reflects all those observers in an opaque way—exceeds its square hold. Something happens in this scene, this moment of releasing our observations, our listening and seeing, registering, and reading, from the hold of "I" into "II," or I'n'I, through this reflective and opaque black square. Antiblackness means to limit yourself (albeit through holding others, blackness, Black lives as limit) and the world, so as to be an individual, as an infinitely enclosed subject that is by itself and separate to its end (and thus ends and must, inevitably, fail). It's like if the artwork, this black square, remains merely something to orient your gaze against, staying separate from the artwork. Won't the artwork remain inaccessible to you? If the music, the color on the canvas, doesn't change you, changes who you are, and is merely some experience for an unchanging thing, then you are already doomed to end, you've already made your end, right here, in black. You end where the art begins, you end where the music starts. But this doubling of *I* (eye; a gaze/view) that happens in this blackness that sticks to you in its glossy reflectiveness is where the end of you becomes the beginning of I, where any I can only be not-me—it is about caring for us beyond a "mine."

Learning to Listen
White Gaze II Black Square is about the question of naming, about what a name does. By naming Blackness, by speaking about Black lives, and studying blackness, we explode the shackles of individuals against each other in an antiblack paradigm for the world that holds blackness in opposition to itself. We must speak about this name, "Black," as the unholdability of names and naming, how whenever we speak about black we must keep studying, we must keep learning about Black lives, about blackness, about us, about the world. Antiblackness is the holding of everything bad inside of this name that is also the attempt of erasing everything with this name "black." But by messing with the word, making it move across death and life, through giving it an uppercase—which denotes Black lives—and then again a lowercase—which denotes blackness as a concept—and by uncovering further uppercases, and changing the lowercase, this name starts moving.[32] It moves, it is alive, in excess of a relation of signs that authorizes them and gives them value, and meaning.

We can only conclude that blackness is not its name, blackness is not its nameability nor its namedness, which is its contextual appearance in antiblackness (as a historically, geographically, biologically contained world)—it cannot be held in a name. This is because blackness must be not its name as it is otherwise simply the demarcation of antiblackness as a categorical self-containedness of the human without black, as unmarked.[33] As I elaborate in chapters 2 and 4, there is this idea of the citizen as unmarked and blackness or race is a mere addition, which actually reifies the unmarkedness of those who are less raced (i.e., those who can inhabit whiteness). Everything that has to do with race, particularly the negatives, including racism, is displaced onto blackness, to the "raced." This is even the case in a social milieu where there are no racial categories, where there is no national account of "racial" identities and no shared discourses around race, racism, Blackness, and so forth. Of course, I am here thinking of Switzerland in particular to make a general observation about the functioning of racism; how the absence of questions of "race" are not the end of racialization and racism. But in the end, Blackness is about more than the question of race.[34] Blackness exceeds the term *race* and it exceeds a system of racism that places some at the top and others at the bottom of the world. The issue is that a conception of some more fundamental—or, let's say, essential—humanness keeps placing blackness as an abject thing of the human, as well as one of the (among other) *instances* of life, while those unmarked remain properly human.[35] Race is placed outside the proper citizen/life in the (negative) ascription of a truth of being, or identity or essence, that being human supposedly is. It is, in other words, through the negative example, the exception to the rule, that those who are not reducible to mere race are made a proper human/subject. Thinking beyond blackness as the loci of mere race, or as absolutely held as only the opposite of whiteness, is necessary to uncover how blackness bespeaks something more than an instance of life—as instances of life have been used as delimitations of time, space, and existence defined against blackness. All these attempts to be without blackness turns blackness into a mere negative of a proper being, a proper sound, a proper world, and so forth. Our celebration of Blackness as and for Black people is not about a necessity of identification; rather, this is about how blackness opens up the question of what may be.

Antiblackness is the erasure of lives on the basis of some kind of fixed thing that we must fit in, that we must be just to be (just to be human, or Swiss, or any other kind of life). That can mean being turned into one among others or an other among ones: how Black lives are made into absolute non-individuals, and at the same time how the very concept of the individual requires a world of no-ones, under the name "black."[36] But it is this meeting, where Black lives, no-ones without names, become named, or where an unname, like "black," becomes a name; where something happens that transmutates the one into a singularity like a black hole, and there a cosmos is revealed, connected by wormholes.

We must conclude that *Black* as a term, as well as a name, cannot be owned. It makes sense since *property* relations—that is, authorized ownership—begins in slavery and is, as a result, entangled in antiblack structures. It makes sense, since Black lives have always been unnamed by an antiblack world. *Black* has been a term naming the negativity of all things valued. The term bespeaks the undesirable other against which all proper names define themselves. Many scholars have uncovered that it is the situation of "Black" to be of social death, and musician, poet, philosopher, and Afrofuturism pioneer Sun Ra theorized such, "Blackness," as appearing under the name of death.[37] In this scene, the term *black* functions as a kind of limit to naming. This is about how ownership over property is claimed with names, with signing. It is signs that express the value of commodity on the basis that these signs reflect the value of the commodity. And such value, the value of the sign, is a result of the relations of exchange and labor that it is enmeshed in. Like a grammar, that is the value of words due to their relations; names/signs are valued because they are of relations. But if Blacks are marked under the name of death, and if a name is only about contractual relations and are not equitable with what they mark (for example, the commodity), then Blackness has no claim to its name. This is why *Black* is an un-name; it is a word that exceeds the definitions of a world that is based on controlling it. Black bespeaks the unnamed, those whose name is erased—like all those Black lives lost to antiblack violence. What happens when the name is stolen away from the grammar that holds it as its opposite? When *black* ceases to be merely the negative backdrop against which names are defined? Like how life is defined against death/black, how beauty is defined against impurity/black. Claiming the name, turning it into an uppercase *Black*, is the demonstrative act of shifting the whole world in relations, into another way of making relations. This naming, this organizing under this un-name, is our meeting in excess of delimitations of belonging and life articulated in and by antiblack paradigms.

Black as name is always only possible as an excess of names and their embeddedness in a grammar, that is, an overarching whole of words, that determines their value. But there is something that opens up, that happens in a transmutative kind of way, to names, words, terms, or signs in general in this naming of *black*. What if words' need for a grammar, or names' need to be enmeshed in relations, is nothing other than the fact that they cannot be in themselves? What if the claim to an authority of naming, to a proper accounting of lives using names or using some kind of measuring stick, breaks down when we look under the hood, so to speak, of naming? When new words come to pass, when we make grammar anew, because of lives who migrate, like *Usländerdütsch* (foreigner-German, see chapter 8) or creolization, then that might just reveal how signs are not sufficient in themselves. Words are entangled in the world and the lives that move in it, that shift what it is. Names come out from the unthought, and more unthought, more questions, come when we name. Naming becomes about lives in relation.[38] Signs cease to be self-sufficient, text exceeds the page, and its sense exceeds relation to only other signs. Signs become simply vehicles, conduits, for music.

There is a common issue here that is of the order of a misunderstanding of names. The idea that blackness is just race, or even just racism, is the idea that the antiblack world makes, or made, it. Antiblackness does not own blackness, does not own *black* as term, nor in any other way. I am pointing here to an issue of antiblackness conceived as that which makes blackness. Blackness becomes reduced to antiblackness itself. If we follow this line of reasoning, then antiblackness had the power for self-creation—*which is the very claim that makes it*. Antiblackness is the claim to absolute authority and knowledge over all life and the world. But if that were possible, then it would not have to make the name of blackness as a marker for that which is its own limit, its own end—which is also to say, antiblackness needs to make itself and in so doing needs to mark its own end, or that which is not authorized to be it. This means also to erase that which it makes its end from within it, dispelling it to a border that can be controlled. It is the claiming of authority over the unauthorized that forms the basis for antiblackness's claim to authority—it is the fabrication and control of its own end, its borders, that belies antiblackness's claim for authority. Antiblackness is the confusion of death/end as a finality of itself, and blackness is that which is in excess of one life by itself with a beginning and an end—an end at the borders of me, of a citizenship, of a nation. Thought as a question of names, it is revealed as a question of who has the right to definition. I don't mean that as merely a question of choosing but rather as a question of the way in which the very right to definition, this authority, is rooted in the erasure of the *conditionality* of names, of signs in a general sense. What happens when those who have been denied the right to definition define, through naming, themselves?

Blackness is togetherness as a coming together to search for a name, in the nonposition of that which is un-nameable as the excess of proper naming.

And such meeting is always already a coming together before knowing that we are coming together (as we don't have a name for such meeting/practice). Such name-search as never concluding, as always only made in organizing or being beside oneself, is a radical refiguring of languages made up of signs (as names) that flip the world on its head. Blackness's generativity of naming, its revealing of name's relational and poetic fabrication, is messing with names in a way that un-names, un-signs—like the *X* that is claimed by Malcolm—name's valuation in exchange for relations.[39]

I engage this sound, *Black Swiss*, as an expression of me as never having been (by) myself, as a way to think beyond me as self-contained. We, Black Swiss, don't really have a name; we never really had a name, and we already spoke about those things that are called in by Blackness. Thus, Blackness is the heading of the un-nameable, a naming beyond names as part of antiblack identification methods. For me, this name "Black" contains and invokes, petitions for all of this care that Afro-women (like Oguntoye et al.) in their organizing gave me as a practice for more than me, for "Me, Myself, and I," for our meeting as the telling and writing of our stories—that is Blackness to me.[40] At the end, the idea for this book came to me only when I heard Black Swiss, and listening had always been the site to hear us. Hearing falls over into the music we make when we sing, and write, of Black lives.

1

Interstitial Listenings I
Charles Uzor's
Bodycam Exhibit 3,
Part I

Interstitial Listening

Music as sounds in excess of what can be known, as sounds that cannot be self-contained, as sounds making unthought relations, is a way to enter the kind of listening that we need and learn for/from Black lives in Switzerland. It is with an attention to how the tools of registering fail in capturing blackness that we can hear how blackness is the continual need to make new instruments—instruments that can make other sounds, new music.[1] The music, and sounds, of Black Switzerland comes to us in the interstices, as the study of the world.

The question of listening to Black lives is a site where we may both problematize an order of registering, or identifying, that continually aims to erase Black lives and blackness and uncover a radical alternate practice of listening that proposes another way of living together. George Floyd's murder in 2020 in the United States has become tied, through the globality of the Black Lives Matter protests, to the case of Mike Ben Peter. Like Floyd, Peter, a thirty-seven-year-old Swiss Nigerian man, was pinned to the ground and killed during a violent arrest by six police officers. According to newspaper reports, this took place because he seemed suspicious and Peter refused a police search on these grounds.[2] What I uncover in chapter 2 is racial profiling: that is the way in which people of color and particularly Black people are identified more often as dangerous and foreign. This is significantly also tied to the claim that Black Swiss do not exist: Because the law states, in Switzerland, that foreigners must carry identification with them at all times—every search implies also that someone's belonging is in question.[3] By disproportionally directing searches and asks for identification to Black lives, it becomes articulated that Black is not properly Swiss. Thus, refusing identification happens for Black Swiss on the grounds of how Blackness is always already marked as foreignness, a question as to whether it should be here. But refusing the search, the ask for identification, takes the form of a preemptive excess of identification, where the order of identification breaks down/open and unveils the attempt to *own identification* by an antiblack world.

While this took place two years earlier than George Floyd, it was the global protests after Floyd's murder that made this case come to the fore due to increased scrutiny about questions of antiblack racism all over the world, also in Switzerland.[4] This scene unraveled, on the one hand, an idea of racism, race, and Blackness as not here, allowing for an erasure of Black lives and antiblack brutalities. On the other hand, it also illustrates how Blackness exceeds the borders of nations and belonging. Furthermore, it highlights how the erasure of antiblackness here can go very much hand in hand with an ascription of antiblackness somewhere else: Black Lives Matter protests at once can be co-opted by an order that continues to displace all questions of race from Switzerland and Europe, but at the same time, protests and care for Black lives can become a way for Black lives to meet and voice themselves. By engaging the globality of George Floyd's case, Black Switzerland, and those fighting for it, shows the shared experiences of Black lives across differences of its particular instantiations.

Bodycam Exhibit

We begin our listening, this interstitial listening, with Charles Uzor and his work *Bodycam Exhibit 3: George Floyd in Memoriam*. A brilliant composer, Uzor was born in 1961 in Udo, Nigeria, and has lived since 1968 in St. Gallen, Switzerland, after fleeing the Biafran war. Studying music from an early age playing the recorder, he turned to the saxophone first at age sixteen, and then the oboe, which coincided with a deeper engagement with composition. Soon after, he completed gymnasium (an equivalent to high school in Switzerland). Uzor completed studies in London, first with a diploma in oboe performance and a master's degree in composition (completed in 1990), followed by a PhD in music composition. His dissertation, "Melody and the Phenomenology of Internal Time-Awareness," was completed in 2005.[5]

The piece *Bodycam Exhibit 3* was written in 2021 and premiered at the acclaimed festival Wien Modern on November 4 of the same year by ensemble oenm.[6] It is the second piece that Uzor wrote in memory of George Floyd. The first one, *8'46" George Floyd in Memoriam*, written in 2020, is discussed in chapter 7.[7] Composed like a chamber concerto with two soloists—the snare drum and the tuba—with an ensemble consisting of a conductor, flute, oboe, bass clarinet, bassoon, horn, trombone, violin, viola, cello, and double bass, its musical score is rather unique: it comprises two components, a set of PDFs and a series of audio files. These audio files are made from the bodycam and bystander footage of George Floyd's murder. The footage used for this piece was also used during Derek Chauvin's trial, who was the police officer who knelt on Floyd's neck for approximately nine minutes, killing him. Uzor redacted for each performer the audio component of the recordings so that they contain only one voice. The musicians are then tasked with imitating the audio recordings for the performance of this work—the music is, to a large extent, based on the imitation of the voices found on the audio recordings of Floyd's murder. To aid memorization, as well as to outline form (and to add further instructions), Uzor provides players with a PDF with the audio transcribed. As a result, every performer has a unique part, and there is no overarching score that displays all the voices in the traditional sense—the conductor has something like an outline of who is silent and at what time, but many of the details regarding what the instruments play remain unspecified in the conductor's part.

In its use of archival materials of Black lynching as basis for a musical score, *Bodycam Exhibit 3* raises a series of questions about the archive, its formation, and its reception or dissemination. It raises the issue of the archive's relation to Black lives and deaths—that is how Black lives are unaccounted for in the archive except as absences, which may or may not mean also their deaths. And, as an extension, it challenges the function of archival engagements along scholarly, political, juridical, and social domains. It thus provides us with an incipit for thinking through the question of the absence of Black lives within Switzerland, an absence evinced most clearly through a lack of archival records in its various forms, such as historical writings and population census.

The questions that *Bodycam Exhibit 3* raises are about listening and identification, as a matter of how things are recounted or left out. What does it mean for the performers to have to listen to this sound-recording in an initial step, followed by practicing the imitation of it, that is also a registering and reperforming of such sounds, and lastly the act of the imitation as a performance of it? What is the role of a memorial in this particular scene, through the medium of music? What is the audience performing by and what is the observer's role in listening to this particular work and, as an extension, as it pertains to the imitated event?

Not only does each musician imitate one voice from the recording, but in the conductor's score, it becomes clear that they are restaging the event, evidenced in a list titled "characters," where each instrument is assigned a person from the actual event (see the following list). Each instrument thus plays out (mainly in sound) one person from the scene, with the exception of the police officers. The musicians that play police officers are doubled. There are two of each police officer from the original scene, and they must search for each other during the performance by walking around. Apart from the police officers and the two soloists—the tuba, which plays out the part of George Floyd, and the snare drum—there are three musicians who play witnesses (oboe, trumpet, violin). They are not obliged to walk around but, according to the score, they can choose to do so as well. All the walking happens only in the second half of the piece, when the police officers start their search.

Instrumentation and "Characters" of *Bodycam Exhibit 3*
Tuba—Floyd
Small Drum (snare drum)—The Unnamed
Double Bass—Kueng 1 (Police officer 1)
Horn—Kueng 2 (Police officer 1)
Cello—Lane 1 (Police officer 2)
Trombone—Lane 2 (Police officer 2)
Bass Clarinet—Chauvin 1 (Police officer 3)
Viola—Chauvin 2 (Police officer 3)
Bassoon—Thao 1 (Police officer 4)
Flute—Thao 2 (Police officer 4)
Oboe—Female 1 (witness)
Trumpet—Male 3 (witness)
Violin—Male 5 (witness)[8]

The positioning of the musicians on stage further reflects the original scene, as well as a critical commentary on the larger societal positioning to it (see figure 1.1). The tuba is slightly offset to stage right, and the snare drum is positioned in the middle and in the back, in a mirrored position to the audience (both of them remain stationary). The snare drum is thus, like the audience, in the position of an observer, as the stage setup suggests. It represents the observer(s) as both individuals and the world, who observed the recordings, as well as the archive where things are recorded for dissemination before being able to listen to or view it. The snare drum's character is titled "the Unnamed," pointing us to another layered meaning of this musical instrument for this performance. This unnamed character, the snare drum with its march and temporally distorted speech-like patterns, evokes the unnamed death of Black lives: the patterns are very self-similar except that there are hiccups, addenda, and augmentations (stretching the speed of musical phrases) or diminutions (compressing the speed). This self-similarity remains there throughout the piece, in the backdrop, lingering and interrupting the flow of the musical narrative. We might hear this as a hint to the way in which antiblackness requires the *putting into* the limit, the border zone, of blackness. As a depiction of how antiblackness creates this backdrop of voices that are not let in, and thus cannot affect the narrative and are outside of the narrative's temporal progression. It seems to tell how all these Black voices that are absent, murdered, and erased keep haunting this antiblack world or this archive that refuses to speak of Black lives.[9]

This spooky relation of Blackness to the archive is at the core of Saidiya Hartman's notion of "critical fabulation," which aims to combine archival material with critical imagining (or what we might call a specific kind of reconstruction) so as to tell stories about those Black voices that cannot be properly accounted for in the archive. This ghostly presence of Blackness, as death, in the world is Hartman's reason for the title of her seminal essay "Venus in Two Acts," where she outlines the concept. As she mentions, these "two acts" correspond to a recognition of "Venus" being the "one who haunts the present" as "disposable" or, we might add, already-disposed "life." And, it is within the "inevitability of repetition" that the dead are returned to us.[10] Hartman explains: "Is it possible to exceed or negotiate the constitutive limits of the archive? By advancing a series of speculative arguments and exploiting the capacities of the subjunctive (a grammatical mood that expresses doubts, wishes, and possibilities), in fashioning a narrative, which is based upon archival research, and by that I mean a critical reading of the archive that mimes the figurative dimensions of history, I intended both to tell an impossible story and to amplify the impossibility of its telling."[11] *Bodycam Exhibit 3* is, in this sense, critical fabulation: it "mimes" and imitates what can

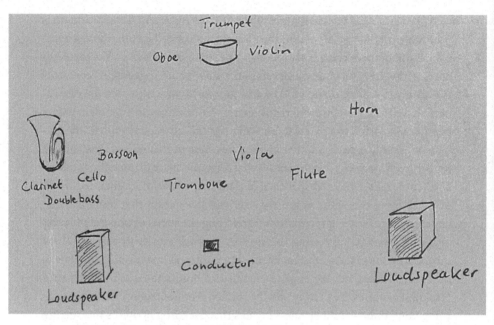

Figure 1.1. Charles Uzor's *Bodycam Exhibit 3* stage plot.

be told, mining and altering in the same turn what is stored in the archive, as well as how to archive.

The problematization of the archive raised by *Bodycam Exhibit 3* happens through engaging it—the archive has to be accessible and must have data. In other words, there is no critical fabulation without any archival material, without any data, without writing that *enters* the archive, nor without history whose (narrative) figuration can be mimed nor without access to it. This (musical) work is only possible after the revealing of the bodycam footage into, and out from, the archive. During the beginning of the trial, the footage from Chauvin's bodycam was not being released. Only after a series of motions by media coalitions did the footage finally get released. First only to individual reporters, and only later was it shared with the public.[12] And Mike Ben Peter—his murder—gets lost in the archive before 2020.

Thus, on the one hand, we must critically reflect on what makes it into the archive, whose life stories can be told. On the other hand, it is not only a question of who can get included in the archive but also about who can be heard and who gets to listen. There are two parts to being able to tell a story: the teller and the listener, the telling and the listening. Both must be able to "speak" of Black lives.[13] Uzor's piece problematizes this very fact, how the seeming inclusion of Black lives in the archive through scenes of death does not mean that Black lives can exist in it. On the contrary, it means that only the death of Black lives, if at all, can be accounted for by a world that seems to define its authority on the exclusion of those voices that haunt it. This is the dilemma Uzor presents us very clearly already with the score in *Bodycam Exhibit 3*. But what happens in this piece is that it goes further than a mere account of another Black death—it unearths how this death bespeaks a life, how there comes, from an interstitial space, between inclusion and erasure, an incessant protest turned into celebration of and for Black lives.

The piece starts with a voice, a voice not recognizable necessarily as a *proper* voice. It starts with the tuba imitating, without any *proper* pitches or other already known musical materials, the first words "George," "George Perry Floyd." We don't recognize the words nor do we have a clear sense of pitch, but Uzor's work nonetheless, retroactively, brings a recognition to the listener of the role of these sounds. While there is no clear pitch, Uzor analyzed the spoken sounds by Floyd, and everyone else, and used this analysis subsequently to derive musical pitches. The first words included in the audio score, played by the tuba, for example, are centered around the musical pitch C. This pitch then becomes part of a collection of pitches for the first entrance of the other instruments. In this sense, only a part of the sound is recorded by analysis—that part that can be turned into a pitch. It is evinced here how Uzor highlights that different ways of recording have also different ways of listening—a computer-aided analysis that *listens* for pitches or a listening that replicates sounds based on the performers' ways of listening. Thus, listening is centered as an act of telling, and it makes possible and impossible certain sounds. In other words, the very act of recording itself has a sound. It is involved in how that which is recounted is told. This does not mean that there is a kind of objective real voice and then a bunch of subjective other ones. Rather, *the way that things are recounted is inextricably linked to the listeners as well as the material and discursive conditions of the structures involved in any account and listening.*

Uzor's work questions the assumptions in conceptualizations of music that presuppose and/or reify contradistinctions of language as proper knowledge and a separate, uncontrollable, fantastical space of music—or another space: that of the absolute unknowableness of noise. By continually shifting the sound recording in the musical retelling of the event of Floyd's murder through imitation, Uzor is asking us to think about *what* is being shifted. Importantly, this is not about erasing; this is about highlighting how an antiblack world erases Blackness, how we must read between the lines to hear

Black lives. It is about listening to that which is left out. Just as the footage of the murder was withheld in an attempt to erase, so does a certain register of recognizing blackness as only death things, as only absent. This is why the protest is that Black lives matter, that (our) Blackness matters, that Black peoples' lives matter—not only their deaths, not only antiblackness and its brutalities, not only life without Black. This shifting of the recording in *Bodycam Exhibit 3* is about highlighting the very *claim* to the narrative over life and blackness that reproduces the erasure of Black lives. But it is also the revealing of how blackness asks us to continue to listen in excess of what can already be heard and what can already be told; to tell of that which is in the condition of erasure so as to learn to listen outside of an authority that claims itself in the opposition to the possibility of hearing anew/otherwise. Learning to listen to sounds unthought, without a claiming of an inside, a space of the properly audible, is to enter an interstitial listening that may weave you into the music of the cosmos.

It is the very claim to having authority over all lives and to knowing what life is—that is also to claim its definition—that creates antiblack brutalities. Think of it this way: the stories of Black Swiss lives cannot be told because Black life cannot exist in Switzerland. There is no one outside of the absolute knowledge of citizenship. Whoever is outside is either from another country or from a country that once existed. *In this equation, there is no Black Swiss.* This also means there is no antiblackness. If there is no account that can be given of Black lives, nor of Black deaths, in a space, then antiblack brutalities cannot exist either. But it goes further. This means that antiblackness itself is not being able to *talk* about it and Black lives. Thus, murdering Black lives is a very physical way of erasing Black lives. And the subsequent erasure of this death can happen on multiple registers. For example, the person could be made into just a person—no one Black. Or the death itself becomes just another failure that simply requires, for example, better training of police officers.

We might think about this fact of antiblackness with the ending of *Bodycam Exhibit 3*. That is the erasure of blackness from within, that can be expressed as an absolute erasure of blackness or the relegating of blackness to the border. The piece follows the real-life event pretty exactly (except for one small deviation). So, at the end, the tuba, who plays Floyd, disappears into silence. The reactions of the other instruments bring to the fore these previously elaborated modalities of antiblackness. The absence of the tuba is first heard with the ensemble playing only imitations of the sound recording—which means there is a lot of movement in the sound—which is followed by slow, static chords or musical pitches held for a long duration. The musical pitches are, in this passage, formally (and semantically) marked by the absence of the tuba, which is also the murder of George Floyd. Their sound is different because it is the first time we hear all the instruments without the tuba playing. We might say that the silence of the tuba contributes to the way in which we hear this moment. At the end, we're left with another snare drum solo passage, with these self-similar patterns played loudly—so loud that we hear it reverberate in the whole room.

In this sense Uzor is asking something of the listener: listeners must confront how they listen. They must ask themselves whether they cover up the absence of the tuba, whether the absence of the tuba is *merely* an absence. Or, alternatively, if all that happens is related to the tuba's absence, and maybe even that there is thus something about the absence of the tuba that is about more than absence as outside of what is here. That is, presence is nothing but absences. This raising of the question of how we listen is what we must confront as listeners when listening to a sound that has a silence in it. Or, in other words, there are different sounds that come to the fore depending on how we listen to the silence(s) in sound.

Steve Martinot and Jared Sexton elaborate on this musical event:

> The supposed secrets of white supremacy get sleuthed in its spectacular displays, in pathology and instrumentality, or pawned off on the figure of the "rogue cop." Each approach to race subordinates it to something that is not race, as if to continue the noble epistemological endeavor of getting to know it better. But what each ends up talking about is that other thing. In the face of this, the left's anti-racism becomes its passion. But its passion gives it away. It signifies the passive acceptance of the idea that race, considered to be either a real property of a person or an imaginary projection, is not essential to the social structure, a system of social meanings and categorisations. It is the same passive apparatus of whiteness that in its mainstream guise actively forgets that it owes its existence to the killing and terrorising of those it racializes for the purpose, expelling them from the human fold in the same gesture of forgetting. It is the passivity of bad faith that tacitly accepts as "what goes without saying" the postulates of white supremacy.[14]

In a discourse marking the moment of another Black lynching as an exception to the rule, acted out by an individual "rogue cop," the very necessity of the forgetting of race—as the putting to the outside of those killed by the police and, in turn, their racialized position as less than human—is articulated.

In other words, it is society's forgetting of race (in their example through the production of a "rogue cop") that performs racism by moving those killed outside of the realm of the human in such forgetting (more on that in chapter 4).

In some ways, this is also a question of musical pitches in this piece. After all, the musical pitches are directly derived from this scene of antiblack violence. And when the instruments play the pitches again without the tuba, with the tuba's silence, the choice of how to listen becomes one also of the real scene. It is in an attempt to distance the musical pitches from that which they imitate, that antiblackness is attempted to be erased. But not only antiblackness, also Black lives. So, Uzor is asking us to think, this piece is asking us to think, about how we listen to the world. The role of us, the listeners, is in making the world into what it is. The snare drum is in the mirror position of the audience, of the listeners, of those who perform the role of the observer during musical performances. In this sense, considering also what character the snare drum plays, it becomes evident that the snare drum is playing both observer and the mirrored opposite of the observer. It plays out the ghostly presence of Black lives erased. Thus, on the one hand, it can demonstrate a listening that does not listen beyond what it can listen to already. For example, if the snare drum is perceived as noise and outside of the *musical* action, since it has no pitches, then it is heard as a lack of music. At the same time, it bespeaks another kind of listening, one that listens with the silences in the sound, with the sound of the room, where the sound is the beauty of silences, like the celebration of Black lives that is also their protest in another register. And hence the snare drum resonates the space, opening space for the telling of our stories. There where its noise becomes protest, where its noise is revealed as silence's overflow into music, it becomes the celebration of ghostly presences that make what we can hear into the sound of music.

This dual question of listening is one as old as the technologies of modern sound recording, and especially listening to such recording technologies. It is a question of how we listen to the world, and in particular how we listen or don't listen to Black lives. Gustavus Stadler's study of early tape-recording consumption in the United States at the turn of the twentieth century elaborates on antiblackness and questions of fidelity, or "realness," of the listened-to recording colliding in the audiences' responses to such new technological experiences. Stadler analyzes the reception and circulation of sound recordings of lynchings of Blacks, which during that time became a popular pastime in the United States. During the same time period, most of the public was also, for the first time, experiencing sound recordings, presented to them outdoors in cities. The recordings of lynchings could not have been carried out at the actual event but would have been restaged in the studio because sound recording technologies weren't good enough yet to record the sound at the scene.[15] Furthermore, these sound recordings, falling in a popular and very common genre of documentary realism that attempted to "capture" sounds of different environments and soundscapes in the world, were produced by the same companies that also produced the earliest musical recordings, many of which were minstrel shows.[16] What Stadler points out is that within the accounts of the listener was an assumption that these recordings sounded *real*. What sounded real had to do with antiblackness: At the time, sound recordings were rather harsh. The technology produced a lot of extra noise. This harshness was seen as befitting of animal voices, screams, and Negro voices rather than the voice of whites.[17] Thus comes to the fore the way in which the question of the realness of sound, and the fidelity of sound recordings, is entangled in a question of what Black should sound like in contradistinction to what a proper voice, one without noisiness and roughness, would sound like.

Stadler moves into another direction when he points out that listening to these scenes of lynching could also become an inspiration for political resistance.[18] In this instance, he is pointing to the manner in which the interaction with a (recording) technology shapes that technology. But that then, in turn, this shaped technology, which carries a particular gaze, can be refused and become a way to reveal what is assumed as proper and how violences of erasure and delimitation accompany such normalization.[19] There is thus a tension between reperformance of antiblackness and a remembrance of Black lives stolen/taken as a form of a wake and/or incipient protest. In the modern digital realm, there is furthermore the question of how and where these recordings are stored, as things remain online in perpetuity.[20] Recordings of murders of Black lives are stored on digital platforms that make a profit from them (regardless of whether directly or indirectly). In this sense, and through the biases in recording technologies against Black lives, these recordings, this archive, can serve (both economically and symbolically) to uphold racist structures.

Recording technologies involved in capturing are not neutral listeners but immersed in systems that are part of the world. Uzor's pitch mapping—that is, his analysis and subsequent rescoring—functions like a distortion of recording technologies' voices; that is, the antiblackness that technology is re-performing through its conditions of existence (in the form of both its making and its support structures). A problematization of the method of consumption and dissemination of these recordings, and thus the role of the listener and the archive as the highways of communication, is part of antiblack structures. This shift is not so much an attempt to nullify the recordings. As they become central to a musical art work, they are indispensable. Rather, *Bodycam Exhibit 3* takes these technologies' antiblackness as grounds for their existence and moves to both a critique and alternate possibilities gained from such critique. It is here that we can hear, in the underground, a protest, a celebration of Black lives, that exceeds the delimitation of the use value or the conditions of these recordings.

Through different ways of listening, Uzor is bringing out the sounds of the technologies themselves. This happens through the highlighting of the interstitial space between the sound of different methods of (technological) listening and sound reproduction. To the aim of this critique, Uzor is engaging in what in the visual domain is called photorealism and in the sonic domain is called *phono*realism. Photorealism describes the practice of repainting images taken by cameras; phonorealism is the same but in sound rather than images (i.e., reproducing *electronic* sound recordings with *acoustic* instruments). The term was coined by Austrian composer Peter Ablinger, for whom phonorealism is less concerned with "literal reproduction itself but precisely [the] border-zone between abstract musical structure and the sudden shift into recognition—the relationship between musical qualities and 'phonorealism': the observation of 'reality' via 'music.'"[21] Recognition here means the recognition of speech, as many of his pieces use sound recordings of spoken text. This also bespeaks again that the question of fidelity is entangled with who can have a voice, who can speak so that it is not abstract but legible sound. Douglas G. Barrett, who has written about Ablinger's work, describes it as such: "The work becomes concerned with the tension between abstraction in musical terms and the potential for music to exist as recorded document. This document, however, does not operate in familiar terms, such as photography or images. It is sound, particularly recorded sound, that serves as the subject of musical observation."[22]

What is this border zone that is between abstract and real, between Black voices and proper speech? Uzor opens the interstitial space behind the border as a control mechanism of what gets to pass, attuning us to the border zone and, in the same turn, changing what it means. The border becomes interstitial listening, becomes a place of possibility to hear Black lives, to hear more than what is accountable in the world that upholds this border as an undesired outside that can only be let in in a controlled manner. It is a controlling that turns whatever gets to come in into something that is a proper image, a correct image, of reality, of that reality that the inside already sees and thus cannot threaten it.

Our ears are opened to the border zone, where antiblackness enforces control over what is real and what is nothing, what is proper and what is mere illegible noise. But with it, we also hear another story, the opening that is the interstitial that lies behind a clearly locatable border zone (which means it is delimitable and ownable through control). This comes to the fore through the snare drum, how it is an undercurrent, or like a mirror, and marks at the same time both a limit and another possible world. It is also evinced in the pitches and the imitation of voice recordings that are the musical materials and score of this piece, through the gaps between these different ways of listening. Lastly, I want to highlight that it is also in the very narrative of the piece, in the very act of telling and hearing this story, that *Bodycam Exhibit 3* opens something up.

Form, in this work, is tied up with the telling of the murder of George Floyd. Thus, in *Bodycam Exhibit 3*, musical form is entangled in questions of narration and narrative and the possibility, or impossibility, of such *for* Black lives. As Frank B. Wilderson III elaborates in *Red, White & Black*, narrative form in film, as in many political theoretical works, moves from equilibrium to disequilibrium back to equilibrium. This is the movement of emancipation—that is, the movement out from antiblack slavery into a freedom or civil society that leaves race and antiblackness behind (see chapter 8). But such narrative form happens on the backdrop of the incapacity of the Negro slave.[23] That is to say that the Negro slave, or Black lives, is not only fundamentally outside of such narrative form—the Negro has no capacity to affect, or be of effect in, such narratives—but also the Negro is that which allows for such form in the first place. This form is the form of erasing Black lives. This narrative form is about gaining control, where disequilibrium is always already gained control over because we must return to equilibrium. It is like the control over a border, where the border is that moment of disequilibrium and already implying the move back to equilibrium.

While this narrative form will become clearer throughout the remainder of the book as a question of Black lives in general, it is also the narrative form of *Bodycam Exhibit 3*. The piece moves from equilibrium to disequilibrium and back, and this is another revealing of a border zone and its underground—disequilibrium. In the piece, this is very obvious when looking at the progression of musical pitches over time, from the beginning to the end of the piece. *Bodycam Exhibit 3* uses musical pitches in a somewhat uncommon manner: each instrument has a collection of pitches for each of the sections of the piece. The instruments then choose freely between those different pitches. That means that while the composer decided on the pitches beforehand, which are derived from the analysis of the sound recordings, the exact pitches that will be played for any performance of the work depends on the choice of the musicians. Over the course of the whole piece, the amount of pitches to choose from changes, from only a few pitches to a lot and back.[24] In this sense, it matches Wilderson's narrative form: from equilibrium, that is low uncertainty, to disequilibrium, that is high uncertainty, and back.

This narrative form is the form of erasure, is the creation of a border zone as controlled, where we always end with equilibrium, and disequilibrium, or uncertainty, is merely a danger that must be controlled by an inside, by a world, that is properly safe, whole, real, and good but without Black lives. But Uzor does something disruptive in *Bodycam Exhibit 3*: one little error exists, one inaccuracy in the retelling of this narrative appears. It is one that explodes the border zone's control; it opens a hidden interstitial listening. It is a jump, a skip in the narrative.[25] It happens right at the moment where the border is made and controlled (this moment of asking for identification). Right at the moment where Floyd is put to the ground by the police, Uzor inserts a skip in the tuba. The tuba skips ahead in the narrative—not away from this moment but into it. But, importantly, it does so by itself, without the other instruments or characters (particularly without the police characters who were the only ones, except of course the tuba and the snare drum, playing before this moment). The tuba jumps to the moment of absolute uncertainty, to this moment where the police must regain control, which is also the moment where control itself is claimed. That is to say, that this moment is where authority over life is claimed, through being able to control Black lives, or disequilibrium. At the same time, this moment is where the need to claim authority comes to the fore and thus can be revealed as not pregiven.

We might make sense of Uzor's choice to make the tuba jump to this moment without having any other instruments, hence also no police, present, through thinking with Mike Ben Peter's case. Initially, there is equilibrium: the world is ordered. Then there is disequilibrium identified: this is when the police demand identification from Black lives. Afterward, the world returns to equilibrium: once there is control over Black lives, in this case in the form of killing, then the danger of uncertainty is in control. (Uncertainty is a danger because it undoes the certainty and thus undermines the authority of this narrative form, or of equilibrium itself.) But by refusing identification, something happens to this moment. It seems that equilibrium starts encroaching on equilibrium in excess of its control. How speaking against policing is before and in excess of policing comes to the fore. It is what policing, here seen as putting Black lives into the ground, fears and needs to claim control over. It is the very demanding of identification that forms a space of proper citizen, or Swiss, that is in contradistinction to those that must be identified and are thus unnamed, unidentified, uncertain, or dangerous (for more on this, see chapter 2). The refusal to show identification is a reopening of the moment right before our blackness becomes marked as the unidentifiable grounds for demanding identification—marked as a controllable danger. By refusing identification into a world order a whole antiblack world is challenged. It is a jump back into the space of uncertainty, or disequilibrium, where the ask does not yet imply an answer. This moment that is really before itself, a moment that is closed off, erased, by the ask that already implies the wrongful presence of Black lives. Right then, as a continual unauthorized reopening of the moment, Black lives identify themselves in excess of identification documents. Black bespeaks the uncertainty of belonging—the rules of the game are upended.

2

Blackness and Black Lives in Switzerland

Speaking Out against Antiblackness

In the face of an *appearance* of a lack of accounts from and of Black Switzerland—that is, a lack of archival material about Black lives in Switzerland—we may engage a growing literature of caseworks that can serve as evidence of antiblackness. In this chapter, I look at cases of antiblackness in Switzerland. The purpose of this is to open the cracks of a world that claims to know it all, to be absolutely *just*, and to be outside of racist inequalities. Through this we can uncover antiblack structuring paradigms, but we can also approach the interstices—that is, the excess of antiblackness.

While Black lives in Switzerland do organize, meet, and discuss, there is (or we might have to say "was," as we are currently in a national reckoning) an absence of not only a nationwide discourse and conception of antiblackness in Switzerland but also of Black Swiss lives.[1] My concern in (re)raising the term *Black Swiss* in this book is to thematize something that was held in an absolute unknown, an outside far away from here, from Switzerland.

There are a variety of reasons why such unthinkability is both a foundation of antiblackness and a basis for national belonging in Switzerland. Firstly, Switzerland is often implicitly assumed to be white while shrouded as without "race."[2] Secondly, Blackness in Switzerland is often considered only a recent phenomenon, mostly argued to stem mainly (or even only) from increased globalization in the last forty years or so, and it becomes subsumed under questions of immigration.[3] Thirdly, Switzerland does not collect any official statistics related to race or racial identities; the only population measurement is regarding immigration backgrounds. These initial three points can be analyzed in connection with antiblackness's need to erase Blackness from sight/within.[4]

Lastly, we must point out that Black lives are themselves highly differentiated: some are different immigrants from across the globe; some grew up in Switzerland but have parents with an immigration background; some came to Switzerland as refugees at an early age and often are strongly separated from their country of origin; and some are Black Swiss whose parents do not have immigration backgrounds or only one parent does. Furthermore, there is difficulty communicating about Black Swissness in general, which also makes it difficult to identify one's own lived experience, and even more difficult to meet with Black Swiss lives. This is not only due to language barriers creating different ways of conversing about Blackness, nor is it reducible to the fact that there are almost no spaces, both physically and otherwise, to engage with questions of antiblackness and Blackness, but we might say that there isn't even a term, such as *Black Swiss*. There are individuals who do use this or related terms, who do fashion spaces for Black lives to meet, but there is no national, or larger societal, agreement, conversation, or, we might say, language to address Black Swiss lives.

Thus, we are already hearing how to think about and work for Black lives requires us to think in excess of what can be properly, discursively, and otherwise made to matter, or sometimes even just thought. In this book I am not so much looking for a proper definition; rather, we must come to grips with attempting to study and think something unthinkable, at least as a properly definable and delimitable category, that is nonetheless very much studied, thought, lived, and sounded. In my view, it is exactly that which is blackness: not a definition, a label, a category, or something that can be reduced to one particular instance. Rather, blackness requires continual study, thinking, practicing, and sounding.

This unthoughtness has something to do with both study and communications, with what discourses can be engaged in (we must also remember there are four different national languages): what can be discussed in the news, what is warranted for study, and what is left out. As Cikuru Batumike notes in his laudable work *Être Noir africain en Suisse*, one of the first contributions in book form to the study of Black lives in Switzerland, there is, surprisingly, not a more national organization or organizing that would allow lobbying for the interests of Black Swiss despite the existence of local groups and a number of Black politicians in Switzerland.[5] However, since the publication of Batumike's book in 2006, there has been a slowly growing set of organizations that lobby for Black Swiss in various domains of life, from fighting for representation in the arts to combating racial profiling.[6]

We can see from laying out the question of where the cases of Blackness in Switzerland are, that inquiring about Blackness requires us to converse with each other. For Black Swiss to meet we need discourses—a language and spaces—while Black Swissness, at the same time, cannot be reduced to particular discourses. As Alexander Weheliye points out, the object of Black studies, blackness, and Black lives cannot be taken as pre-given but rather articulated with theorizing, thinking, and writing about Blackness.[7] Thus, thinking about and studying blackness and Black lives is also an act of self-definition—it is as much a practice that is alive as it is an inquiry into the world. Nonetheless, blackness is already there, before black study; we can hear this through antiblackness, but there are also Black lives beyond what is commonly made audible. Thus, the articulating of the object of study is not so much to be understood as an agential will to creation but rather some kind of unlocatable agency that exceeds knowledge. There is blackness there wherever antiblackness goes because antiblackness attempts to separate itself from blackness. In other words, blackness is always there in spite of, in excess of, and beyond the purview of any antiblack measurements and delimitations. Hence, the writing about Black Swissness, even though an incursion into language, into thought (in Switzerland), is not an imaginary projection separatable from the world but engages lives.

What does it mean to not be able to speak about Black lives in Switzerland? What does it mean to have no way to think Black Swissness? What does it mean for Blackness to *not exist* inside Switzerland, except as an absence or temporary presence? These kinds of questions bespeak the problem of this mythical absence that is a lack of thought given to Black lives and blackness in Switzerland.[8] And these kinds of questions cannot even be asked before this name, this term, this sound of "Black Switzerland." Let me begin with a case, one of many that is an example of an everyday occurrence and a fundamental structuring principle of antiblack delimitations of belonging, as the erasure of Black lives from here.

In 2015, Mohamed Wa Baile, a Black man with Swiss citizenship and immigration background, refused to present his identification to the police in the city of Zurich. The police officer argued that he asked for identification on the basis that Wa Baile *seemed* suspicious.[9] This suspicion resulted from Wa Baile's behavior—this *behavior* was to not look at the police officer. But this perceived suspicion immediately calls into play a system of policing of Switzerland's borders, of who should be here and who should not. This is, as briefly noted in chapter 1, because in Switzerland, asking for identification engages a law that states that only foreigners, not Swiss, must carry identification on their person at all times.[10] Thus, the practice of asking for identification holds a specific kind of performative relation to racial discrimination through the ascription of foreignness in those chosen for control. In this scene, it is clear that asking for identification is not a neutral act; it has itself already a meaning and *questions* proper belonging. That is not to say that every ask for identification is about whether you belong here or not, but rather that the act itself always carries this question with it. In fact, the very asking for identification implies always already identifying those asked for identification as suspicious, as possibly being not Swiss, and thus a foreign and possibly dangerous element. This questioning of proper belonging through identification of those who do not look like they are from here, or who seem suspicious, is itself racializing: it claims that those who are not asked for identification do not all pose a danger. The bias against Black lives and people of color is in the erasure of the way in which those whose belonging is not questioned are always made into the image of Swissness.

As a matter of fact, the erasure of racialization is fundamental to racism's working. For the claim over racism can only be raised as a juridical tool against those who become profiled due to their race. This in turn also means that racialization itself is protected by the judicial system. In other words, if police profile someone due to their phenotypical appearance, then they are, in the eyes of the judicial system, not acting racist because racism does not exist. But when someone opposes a racializing practice, discourse, or actions, they are in the wrong in the eyes of courts because they practice a kind of discrimination by calling out racism.[11] Because (antiblack) racism is reduced to a question of individuals and their willful wrongdoing—a question of discrimination only—calling out racism can itself be turned into the same thing: an act of discrimination, or at least defamation.[12] That does not mean that there are not cases where individuals should face consequences for antiblack or racist rhetoric. Rather, it means that the way that racism is conceived actively influences how such laws are applied. It is under the guise of a race-free space that antiblackness is continually articulated, that racialization takes place, and that nothing can be done against it because of the erasure of antiblackness and questions of race and racism.[13]

Wa Baile's refusal to show identification is thus a protest against being labeled "foreign" on the basis of his appearance, of him seeming suspicious, of him representing the image of possible danger. For Wa Baile, it was clear that it was because he does not look white, because he is Black, that he was stopped, and thus he refused showing his identification. Upon his refusal, Wa Baile was arrested for not following the police officer's orders. During the subsequent trial, the defense attempted to explain the defendant's position and structural racism—the bias against nonwhites in policing. They elaborated on what structural racism is and how it influences police officers' decisions and their biases for who gets asked for identification.[14] But racism is dealt with in a rather peculiar way in Switzerland's courtrooms, as this case evinces. Despite the fact that the Swiss court system recognizes racism as a problem—evidently Switzerland is part of an international agreement to combat racism[15]—there is nonetheless little chance that courts will rule in favor of instances of racism.[16] There is an underlying idea that racism is obvious, that it is reducible to an observable and explicit action perpetrated by individuals against a group of people, rather than a larger societal issue. This carries with it a number of problems—most obviously that systemic antiblack racism is also present in the courtrooms that judge on such matters themselves. But this is a difficult claim to make because if racism is reduced to a discriminatory act that is easily observable, then how could a more complicated systemic account of racism—that is, exactly the very biases that appear when policing against a group of people—be thematized since data collection is almost impossible? Not only is data collection difficult, even having a discourse proves almost impossible. It is exactly under this same kind of conception of racism that the modern prison system in the United States came to flourish, that killings of Black lives by police continued to grow.[17] The erasure of racism is coupled with the erasure of race and the racialized, turning race and racism into something concerning only the individual and as pathological.[18] What is evident is that the apparent lack of factual data for racism is itself not necessarily due to a lack of cases, but rather because of the biases that antiblackness itself instantiates. It is about what gets counted, what can be discussed, and what can be studied, which is also to say what is inhibited from being discussed or studied.

By the end of the trial, it comes to the fore that Wa Baile seems to have even convinced the judge somewhat of the fact that there might indeed be a possible (structural) issue of racism here. While the judge defends the police by pointing out that they receive antidiscrimination training, he also acknowledges Wa Baile's account of the events. First, the judge states that he believes Wa Baile's account but that it is not verifiable and furthermore is deemed "irrelevant" to the proceedings and the ruling.[19] And then, a little later, he states that Wa Baile should be lauded for standing up against injustices in society, but that he should nonetheless follow the orders of the police. Thus, while the judge hears the defendant's argumentation as believable, it was nonetheless inapplicable to the court proceedings because his account was deemed purely subjective and outside of the questions that this civil society can deal with (or wants to, at this moment).[20] The court ruled against Wa Baile because there is no way to think and speak about, and of course no legal tool to deal with, this question.

This is a question of an inability to bring the very sound of Black lives to the ear of not only this court system but also society at large. The officials involved, just like the systems in play, cannot properly subscribe to systemic antiblack racism while there is nonetheless something like an unspoken knowledge of the veracity of Wa Baile's claims. Not only is this reinforced and caused by a mere absence of legal tools, but the very absence of legal tools is coupled with, reinforcing, as well as coming out of, an absence of thought, as a set of reenforced behaviors and actions taken by individuals in the room, in society. But we can also hear how Wa Baile's protest not only uncovers structuring principles of behavior and thought, on both an individual and a societal register, but also bespeaks a radical departure from a world of antiblackness and the agencies within it. It also bespeaks such through the very holding up of a mirror that shows the conditions of the world to the world, that also, in the same turn, hints at something that always exceeds the conditions of the world in them (more on that later).

The statements made by the judge (and the defense) throughout the court proceedings were entrenched in what we can call "color-blindness"—erasing all questions of race and racism.[21] There are three modalities that can be readily identified in this case, that correspond to the main modalities of color-blindness.[22] One, the court invokes notions of equality among all citizens—using abstract ideas rooted in liberalism.[23] Two, the court de-thematizes racism—explaining the effects of racism away by using other parameters.[24] Lastly, the court sympathizes with the problem of racism but only so long as it is a subjective and individual question.[25] By not recognizing racism as a structural problem, making it basically an individual's fault or feelings, it is erased from Switzerland itself.[26] These modalities prove how the very discourse and space within which this case took place are incapable of speaking about racism exactly because of racism as color-blindness. The case concludes with the judge making a comment after his verdict, which is that the defendant is found guilty and has to pay a fine. He states that Wa Baile's combating of racism was admirable, and he should continue to voice against it, while at the same time he should follow police officers' (and the law's) instructions.[27] The defendant laughs after this bizarre statement by the judge, and the judge reprimands him. Racism can be discussed, but only while maintaining the current order, and to speak up against it, against racism, is to trouble waters that shall not be stirred.[28]

Wa Baile continues to fight antiblackness in Switzerland: he coedited a book on this case, bringing this case, with the help of the Alliance for Racial Profiling, all the way to the European Humanitarian Court.[29] He exemplifies that Black Switzerland is making sounds, even when made inaudible, and has a sound beyond what is claimed to exist and as audible. There are other examples of people working for and studying Black lives in Switzerland, in spite of this seeming lack of data, this inability to register or hear, Black lives in Switzerland. One case, or example, is given to us by Carmel Fröhlicher-Stines and Kelechi Monika Mennel. Fröhlicher-Stines has been doing groundbreaking work on matters of Black Swiss and racism in Switzerland since the late 1980s with the group Women of Black Heritage, which she founded with Zeedah Meierhofer-Mangeli in Zurich.[30] Fröhlicher-Stines, a psychologist and literary and linguistic scholar, was born in Haiti, studied in New York City and Switzerland, and has lived in Zurich, Switzerland, since 1971.[31] Fröhlicher-Stines and Mennel's study, *Schwarze Menschen in der Schweiz* (Black People in Switzerland), is a qualitative sociological analysis where they interview Black Swiss specifically to understand Switzerland's antiblack racism. To this aim, that is, to collect data on Black lives and antiblackness in Switzerland, they need to wager some sort of definition of Blackness in Switzerland in order to state, and decide on, whom they would interview. They define Blackness in Switzerland as people who both have black skin and have African heritage (they use the German *Erbe*, which also denotes inheritance, bequest, and legacy).[32] Hence, Fröhlicher-Stines and Mennel posit two ways to register Black people: one, through skin, and two, through ancestry. These two registers are, of course, common ways of thinking and especially registering Blackness, and they are also the common methodologies of racialization in effect in the world. At the same time, these descriptors point to a series of problematics that become amplified in the absence of a name for Black Swissness, as fundamental questions about blackness and racialization.

Fröhlicher-Stines and Mennel open their study of Black lives through reaching out to Black lives—studying Black lives requires meeting them, having them tell their stories. The question of registering Black lives is thus a question of thinking about and fighting for them. At the same time, thinking about Black lives, in the case of Switzerland, points to the fact that it is also inherently a question of how to listen to lives—it requires communal discoursing. Fundamentally, it requires society to ask itself how it thinks, identifies, and deals with lives and is as such a question of self-reflection as a conversation with each other and beyond oneself. When questioning someone's identity so as to question their belonging here because they appear Black (or closer to Black), and in turn articulate a certain normativity of who should live here, border policing is revealed as a question of the inverse of this asking about lives—it closes life off, it closes off what lives (here) can be. This kind of registering, where any person who is not a citizen must be identifiable, is a claim to authorized knowledge about what, or who, should live here and who should not.

Policing Borders

Usually when people think of the control of Black lives, black skin becomes equated with an ability to see them. Black skin becomes a clear marker, a way to register, identify, and control Blacks. It was of fundamental importance in early theorizations of race, and it is an overdetermined mode of registering blackness in both the biological and beyond.[33] As the biological foundations of race have become problematized and demonstrated to be ill-founded, these markers have nonetheless retained their importance. Racist ideas are ascribed to biological markers and, at the same time, the study of biological "markers" is shaped by racist thinking.[34] Interestingly, it seems that antiblackness has never been particularly interested in seeing Black skin, despite the fact that such skin is used to claim Blacks' inferiority. Rather, such happens through the unseeing of Black skin, as a way to control blackness. The need to control blackness and Black lives, to control the definition of life and to delimit it, requires the unseeing of blackness, an unseeing enacted through skin color. The racist controlling and registering of Blacks are effectuated through an erasure of Black skin, or blackness on skin, itself. It is the very need to be without race, without blackness, and skin that could be black, which also means the possibility of being or becoming Black, that requires the erasure of black skin, which is also why skin is so important. Skin becomes a container of blackness: it is a marker, a border, that holds blackness off from those who are life proper—those whose humanity is not questionable. This is also why I must protest: Black is beautiful. I protest not to fetishize Black phenotypes, another form of antiblackness, but rather as a love for the people.[35]

Modern technology evinces this and its entanglement with border control. Pseudoscientific bases for racist delimitations of life are continually reproduced in modern technologies, such as face recognition and iris scanning, which are used in passport control at airports. Blackness's place in relation to citizenship and nation-states is repeatedly questioned through invisibility and hypervisibility. These modern technological apparatuses have biases against Black bodies and manifest themselves through being attuned toward white skin, eyes, and faces. If there are biases that favor white bodies in video and camera technologies, iris scan technologies,[36] and within AI technologies more broadly,[37] then these come from the practices and values of those making them as well as the structures *behind* these makers.[38] By structures I mean the social, economic, institutional, national, and historical conditions of the makers and the technologies. The biases in the technologies thus in turn also reveal biases that are present where such technologies are not in use (i.e., they reveal clearly the biases of the structures involved in their making). If there are biases to be found in biometric technologies to control borders, then, of course, those systems behind their making must also express, or wrestle with, the same biases. Technologies reveal racial profiling as systemic.

In Switzerland, for example, where there is a considerable lack of data as it regards incidents and questions of racial profiling, such lack of data is accompanied by an understanding of racism as exactly not systemic. Evinced by these technologies is that a deliberate lack of certain kinds of data is racism's modus operandi—how these technologies' biases appear due to a lack of being able to see Black lives, exemplified in the way in which statistics are collected around immigration. While there are indeed no measurements of race or racial identity in Switzerland, there are measurements of immigration—measured according to countries of origins. In this sense, these statistics serve to manage immigration. At the same time that these kinds of statistics articulate an inside that is unmarked, they also relay the question of ethnicity or race, and with such also racism, to the borders. Because border policing is a question that cannot be separated from questions of antiblackness, as evinced in these technological biases as well as the impossibility to thematize such biases inside, this kind of data collecting supports an idea of race and racism as far away from the inside of the nation.[39] There is a direct correlation between choices of how populations are measured and conceptions of citizenship. It is also a matter of national narrative and self-definition. The racial becomes relegated to the borders and, in turn, the nation becomes race free, supported by its own biased way of measuring and collecting data.

This becomes very clear when scrutiny is placed on the practices of border policing. In 2018, Mattea Meyer, a Swiss politician and member of the National Council, petitioned for the disclosure of those criteria used by border patrols to choose whom to control at border crossings, specifically in relation to how ethnicity and skin color are used to inform policing decisions.[40] The council denied the request, stating that they take a multitude of factors into account and that they do not base their decisions on one criterion by itself but that they also will not disregard anyone because of *a* criterion.[41] This means that if there is a practice of identifying and policing border-crossing populations by skin color, it is justified as long as there is previous data to support a claim that a certain skin color, or "ethnic group," can be grouped into the category of possible danger.[42] Thus, the very way that data is collected reinforces continual biased policing, itself the very mechanism of collecting data that justifies it. They furthermore claim that their patrols receive training as it regards both racism and discrimination and imply that it couldn't possibly be racism.[43] In these statements, a discourse covered in color-blindness comes to the fore—discourses of color-blindness circumvent the discussion of race as a structuring condition for the interaction between individuals and border patrol. In their insistence on only using ethnic markers without being racist, the border patrol thus overtly minimizes racism and (somewhat) covertly naturalizes racism's effects.

That this is not only a question of a racist set of actors alone but rather demonstrates a structural problem can be further evinced in the discourses around a renowned and scandalous ad for the right-wing political party the Swiss People Party (SVP). In 2007 they created an ad for legislation, as well as for elections, which read "*Sicherheit Schaffen*" (make safe) and depicted three white sheep inside the Swiss flag kicking a black sheep out of the red Swiss space (see figure 2.1).[44] It is obviously xenophobic imagery that, at the same time, reveals the racialization of national space and safety. The inside of the space of Switzerland is not only unmarked but also white and the outside not only marked but black.[45]

The discourses around this ad, not only from the party but also from other political parties, associations, and organizations in response, reveal much about the way in which antiblackness works in Switzerland. Many respondents critical of the ad evoked principles very dear to Switzerland, such as its

Figure 2.1. SVP "Sicherheit Schaffen" sheep campaign poster.

apparent tradition of openness toward others (related to Switzerland's neutral status). This response was conceived as an affirmation of the diversity of the Swiss community and its historical heritage, as the response by the Geneva local parliament demonstrates, who fashioned a counter ad depicting twenty-three different-looking sheep, symbolizing the twenty-three cantons that make up Switzerland.[46] But while this could seem to be against the initial ad, none of the responses addressed the fact that the sheep was black and the ones inside white. They turned the discourse into one about a question of accepting different cultures, or ethnicities, playing into the discourse of color-blindness upon which the sheep ad relies in the first place. The question of the fact that the sheep that is thrown out is black is circumvented by a discourse that sees it not as a question of antiblackness but as a question of a diversity of origins, cultures, or heritage. In turn, it also plays out a discourse of color-blindness that "minimizes" racism as a question of national belonging. As Swiss scholar Noémi Michel points out, these responses played right into the

SVP's subsequent defenses. Ultimately, no action was taken, legally or otherwise, against this incident because of this.⁴⁷

Telling is an SVP representative's defense for the ad:

> Opponents to SVP, you do *confound a saying with so-called racism*! . . . Your hate towards the SVP and its black sheep blinds you and is so extreme that we should call for the interdiction of the use of the word "black" from our French language. Recently I was at the post office and the officer was asking the client standing before me to sign a paper with a black pen, because this would be better than blue. She blushed once she realized that the man was black. *Was she racist?* . . . A vine grower won't be able to name his wine *Pinot noir* and if someone drinks too much, it will be forbidden to say that this person is *noir* [black]. . . . *Is this racism?* No, all these images do not constitute racism. But *you are exerting racism* towards the SVP when you declare that 25% of the population that voted for this party is racist.⁴⁸

By attempting to make the question of the sheep's color not about race, the SVP was circumventing what actually is articulated in the poster—namely, a question of borders and national belonging. Michel concludes that through this, the SVP is able to make racism itself a question of those who see racism:

> Those who pronounce explicit racial references are the ones who cause racism, the ones who "provoke" and "exert hate" against the SVP and its supporters. In order to negate the sheep poster's association with racism, the SVP's defense draws upon the prevailing regime of racelessness marked by the taboo of explicit verbal reference to race. On the other hand, without denying the racism of the poster, the Geneva's motion diverts the controversy towards "exclusion" and the denial of foreigners' contributions. Moreover, it proposes an evasive interpretation of the figure of the sheep, affirming, for instance, that "*each* person can identify with the image of the black sheep."⁴⁹

In this incident, a belief in Swiss values of openness, inclusivity, and neutrality displaces any discourses around race, racism, and antiblackness. Both the evasion of whiteness in the form of appearances of multiculturalism as well as the claim that racism is nonhierarchical (for example, the SVP claimed that color has nothing to do with it and that calling someone racist is racist, i.e., that racism is not only not systemic and individual but also that anyone can be racist against anyone) reflect color-blind modalities that at the same time reperform antiblackness. Her analysis of political discourses within Switzerland points to a practice of border control as one, firstly, involved in racialization and, secondly, a certain political performativity of Swissness that is rooted in the contradistinction to blackness.

The second point can further be clarified by thinking through (Swiss) nationality's investment in ideals and how those ideals are ascribed onto cultural practices as well as members (or negatively to nonmembers) of the nation in a racialized manner. This comes to the fore in Michel's brilliant analysis of the counter by Geneva representatives that claimed, in a first step, a Swiss valence on nonexclusion and, in a second step, marked Switzerland as already diverse or multicultural. In this comes to the fore how color-blindness displaces diversity into the past. By claiming that Switzerland is diverse on the basis of it having been formed by "migrants" and "foreigners," racism as a question is displaced by time, erased from being of question in the present.[50] But this time is a temporal narrative of progressing away from race. This means, to put it in somewhat general terms, that the erasure of racism now happens through the ascription of "something like racism" having happened then. Racism and race are again displaced to the borders of the nation, not only spatially but also temporally. Lastly, as the Geneva council suggests themselves, this proposes that "each person can identify with the image of the black sheep,"[51] which, on the one hand, as Michel suggests, reinvokes a "regime of racelessness," avoiding thematizing race.[52] On the other hand, the erasure of race here, through the claim that anyone can identify with the black sheep, also points to how whiteness must erase itself. Whiteness erases itself as a normalized subject (citizen and/or human life).[53]

The displacing of the question of race also happens through the evoking of culture and a discourse of good citizens and the foreigners' contested goodness. In the right-wing discourse, this comes more readily to the fore, where it becomes evinced that a criterion for joining Switzerland is having to prove that one is *good*. Let us remind ourselves that color-blindness displaces questions of racism to other parameters, particularly culture. The SVP correspondent states that one should welcome foreigners, but only if they are good for the community.[54] In this statement, it is implied that there are desirable and undesirable foreigners, some that provide a *profit* and others that do not, that might even be negative, a danger, for the *community*. Thus, the value of a foreigner can be measured according to their contribution and participation in the community. The immediate question unarticulated is what the community *is*. Of course, it is implied that such community is the Swiss community, but it is unclear as to what constitutes it. In other words, the community itself, that which is good, emerges only by questioning the foreigner and defining thus what is bad, or dangerous, to it.

There is implied in this statement that there is a prefixed and established community that is Swiss and holds *good* values. This also means that all Swiss are carrying in them such a goodness or that they carry it for the community. At the same time, this carrying is unquestionably given to those who are of the community and in question for those who are not. This has two consequences: one, the community is marked by a common essential trait, and two, those who are foreigners are in an impure position, as it is not clear yet *if* they bring profits or danger. Thus, it is not only that foreigners are questionable in their status of belonging, but they also are questionable with regards to whether they hold that which makes someone a proper citizen, which *fixes* citizenship and community. The crux of this example is that such a position of scrutiny, the position of possible danger of which the foreigner is exemplary, is a racialized position (and is colored in these metaphors in black). At the same time, unquestioned, unmarked, and, in this example, white are those that own and are of the space of Switzerland. Furthermore, they are made such in pointing out those who pose a danger. Thus, the question of cultural racism is one that displaces the question of race by engaging in discourses under the heading of culture that are racializing/racialized and displaces the very question of race away from that inside that is the unchanging community onto its borders, there where the potentially dangerous and unidentifiables are.

Controlling Kinship

What we have seen in these examples is that Switzerland as a nation relies heavily on a discourse rooted in color-blindness underneath which brews systemic racism evidentiary in the inability to discuss racism, and in the practices that do not exclude skin color as a marker but erase such measurement at the same time. The erasure of race, racism, and Black lives works through the creation of border zones, places that are at a distance, and absolutely separatable, from inside Switzerland. These contested spaces are where blackness is held, held off from coming inside, and also purified, so that whatever does get taken in can never be black. This might all seem somewhat abstract, but let's remember that this abstraction is physically enacted—from the example of police profiling to biases in the judicial system. There is another border, another space that requires control, in a world that is antiblack—namely, kinship. It is a question of who gets to enter into relations and who doesn't, how the nation continually exerts control over reproduction (see chapter 6). It is a question of how citizenship is not only through an idea of some kind of inherent symbolic trait, as being beneficial to the community or as some kind of essentialized Swissness, but also through biology, demarcated and controlled.

In Switzerland, there exists a rather recent *method* of border control alongside lines of kin. While recent, it is an outgrowth of earlier immigration measures and conceptions of Swissness. Swiss citizenship is rooted in jus sanguinis—citizenship is conferred through kinship/blood rather than through being born within Switzerland.[55] Before 2006, Swissness was passed on only through the mother, or the father when he married the mother of the child. After 2006, the child of the father could become Swiss if it could prove this blood relation.[56] In 2008, the same year Switzerland joined the Schengen Agreement, which allows for free movement between states of the Schengen Area, a new law was passed to improve the management of immigration. Thus, while there appears an improvement in free movement, such freedom is only for some, and particularly not for others. Article 97a, titled "Circumvention of the Legislation on Foreign Nationals," states that "the civil registrar shall not consider a request for marriage if the bride or groom clearly has no intention of living together but wishes to circumvent the provisions on the admission and residence of foreign nationals."[57] This article is intended to curb "abusive union claims," a fear of what might be termed "sham marriages" that merely serve to gain citizenship.[58]

Researcher Anne Lavanchy conducted fieldwork between November 2009 and January 2011, observing civil registrar officials under the guise that she was interested in the ways registrars test for the validity of marriages between Swiss and non-Swiss. At the end of her fieldwork, she interviewed twenty-three unsuspecting officers of the registrar.[59] Lavanchy particularly collected data as to how these agents conducted interviews, interviews intended for them to be able to determine when a couple's relationship was not genuine. Lavanchy describes the process:

> These [interview processes] are very similar to criminal hearings in their form, as both fiancés are heard separately, on similar topics, and their answers are crosschecked in order to detect inconsistencies, which should signal faults and abuses.
>
> Representations of nationality are at the core of these measures: the claimants' nationality determines the kind of documents that they must provide and also the degree of reliability of the documents: countries listed by the Foreign Affair Ministry as "risky" are believed to deliver

unreliable documents, and their nationals are frequently suspected of fraudulent intentions, as if the unreliable character of "their" national administration automatically implicates them.[60]

What comes to the fore in her account is how foreigners are potential dangers, a riskiness that depends on the type of nationality the person has. The foreigner, as a category itself, is a contested space, a kind of border zone. This border zone engages the question of proper fit for becoming Swiss, or of someone's children becoming Swiss, at some point in the future, because of the laws governing how citizenship is passed through kinship. The foreigner as a border zone, as a holding off of the bad from the inside, is a question of kinship and proper blood relations. The question of fitness for marriage is entangled in the question of receiving citizenship, in particular for the children of the couple, through the laws of governing the attainment of Swiss citizenship. While marriage also opens up a possible path toward citizenship for the parents, the scrutiny of proper fit is bound up with questions of filiation.

That this is a question of kinship is further evinced by the fact that the foreigner, as a category of inquiry, is coupled with the category of the "mixed": with a person who has multiple nationalities or *national origins*. Lavanchy observes that such questions of mixedness are, additionally, not reducible to a mere question of different passports but rather tied into questions of ethnicity and race. Lavanchy explains: "At first sight, mixedness seems to refer to 'bi-national marriages' as marriages between Swiss are less likely to be problematic. But nationality is here explicitly linked to 'physical discrepancies' and the idea that in real couples, people should 'match together.'"[61] This becomes very evident in one example that Lavanchy provides of a couple comprising a man who was Philippine Swiss, who was adopted, and a woman who was Vietnamese. As the registrar explains: "I've got here a Swiss man, born '85, with a lady from the Philippines. But his mother is from the Philippines so I think it is normal, he's got a link to the country. . . . This one was a very nice couple. She is from Vietnam, he is Swiss, but an adoptee; in fact he is also from there. It is normal for him to look for his origin, it is quite understandable."[62]

What this case demonstrates is how discourses of Swiss nationality are entrenched in questions of origins and kinship and with, ultimately, racialization. As Lavanchy poignantly observes, the Swiss man never properly loses his Filipino heritage and is hence always not fully Swiss. Secondly, the argument implies that a relationship between a "proper" Swiss—that is, a Swiss without an immigration background—and a Vietnamese (or someone from "there," some kind of abstract region of geographic space that seems to be more about race than about nationalities) could be registered as besides the norm. In this case, the couple is not properly a *mixed* couple to the registrar because the man is not really fully Swiss and is marrying someone who is "like him." Thus, this scene of border control along lines of kin is marked by the undercurrent of the mixed, where mixedness is simply ascribed to him and thus there is no danger of *mixing blood* apparent in this scene—he never really became properly Swiss.

Switzerland's registrar conducts interviews under the auspices of racelessness, and to remain so, this figure of the mixed and the foreigner is created as a border zone to (and thus separatable from) it, marking itself—that is, the sovereign national space—in the same process as pure. It is thus evinced here that notions of color-blindness and of a kind of multiculturalism, as the idea of being mixed and of multiple origins, work in tandem to erase all questions of race, racism, and blackness from within the nation and the citizen.

Borders are these spaces where race and proper belonging, as a question of nationality but also kinship, are articulated. Here, the border is the foreigner and the mixed, which can be both a person or a relation. Mixedness or miscegenation—that is, the mixing of "proper bloodlines"—is revealed as "a precondition of racial categorization" as well as the formation of proper citizens.[63] Rather than overcoming race in these scenes of border policing, the notion of mixed-race couples, or people, founds it.[64] There is in the idea of mixing bloodlines, or blood, an assumption of an already previously pure, unmixed blood. Take again the previous example: anti-miscegenation is practiced by way of the presence of miscegenation—the mixed Swiss man can be with the Filipino woman because he is really also Filipino. In other words, through evoking originary kinship, firstly, mixedness is demonstrated, which also makes a white person and a Philippine person as other, and, secondly, the erasure of impure blood, or race, is achieved since there is no interracial relationship present. We could also say it this way: the need to inquire about mixedness presupposes purity and a claim of such for those that *properly* belong.[65]

By promoting mixedness, a diversity of origins, or multiculturalism, anti-blackness is thus not necessarily circumvented. Rather, such can very much be the root of it. By projecting mixedness into the past, into the foreigner, the purity of the here is reaffirmed. These "mixed" people become a kind of border zone between the inside and a dangerous outside that could corrupt the inside, the nation, and its citizens. Thus, racial discourse does not only not disappear by engaging notions of a diverse and multicultural community, but they serve the erasure of race from those who are proper, and also erase the question of race and racism itself. Race is thus not only about affirming a kind of purity inside. Rather, this purity is made by articulating an impurity that continually protects this purity by being locatable at a distance, a border that can be managed and controlled.[66] And this purity is itself upheld by an erasure of the discourses, and everything else, that connotes race or racism from within this pure space.

Seeing the couple as befitting due to its sameness, based on the mixedness of the Swiss, is the reaffirmation and performance of whiteness as a normalization of the proper citizen. Mixedness, or multiculturalism or mixedness as a figure of multiculturalism, is not only not raceless, but it seems also to be indebted, as well as conducive, to color-blindness. By making the Swiss man mixed, and thus also not properly, or fully, Swiss, Swissness becomes both a pure and raceless category that places questions of race onto the foreigner and those who are mixed. In the example of the registrar is articulated, under the heading of protecting national borders, a racialized understanding of citizenship. A decision as to whether someone matches with someone else is made (so it is claimed) without the question of race present, which ends up in discourses of national origin as a measure of potential future citizens. Color-blindness and certain modalities of multiculturalism are two sides of the same coin: an erasure of race and racism, but also the ascription of race onto others. This question of origins then, in turn, professes a multicultural society where different ethnicities can live together, hiding racialization in plain sight. In this, an unspoken purity of Swissness is reified. They are those that are not racialized; they are neutral or pure, and they are so in contradistinction to those with race, those who are impure, those whose goodness and belonging is under scrutiny. Thus, a hierarchical and racialized notion of citizenship and nationality is articulated on the borders of the nation/citizen, in the geographical as well as the biological.

The Color Line

This interplay between racialization, race, and national borders is further made evident by thinking through the category of the "foreigner" (*Ausländer*) and how it is engaged, traversed, transgressed, and managed. Both Lavanchy as well as Viviane Cretton's work can be seen as attempting to uncover this blanketing signifier of the foreigner and demonstrate how such category is racialized.[67] Cretton unpacks the category theoretically with more scrutiny under the heading of nationalities. As she poignantly observes, not all grouped under the heading "foreigners" have the same social standing or possibility of shedding the status of being a foreigner. Particularly, she notices how the category of the "African" is placed at the bottom of the social scale and how others, who fall within the category of foreigners present or past, perform this kind of move away from blackness.[68] Cretton points out how stereotypes in relation to Blackness are evoked by other nonwhites with immigration status, to separate themselves from the "new" foreigner, or immigrant, and to align themselves with those from here (i.e., Swiss).[69] As she notes, this is how Switzerland, and belonging to Switzerland, marks itself as "white," and, we might say, antiblack.[70] It does not matter that within the category of the foreigner, there are also people who are white. They likewise become properly white when they shed their foreigner status, their impure position. On the other hand, what the examples of the nonwhite Swiss immigrants prove is that to become Swiss is to renounce blackness. For Cretton, this is in some ways a question of "new" versus "old" foreigner: "In Switzerland, like in Europe and elsewhere, the old and the new im/migrants (the Africans being the latecomers to Switzerland) are overtaken by racist relationships and therefore compete for greater social recognition."[71] While an important observation, I propose to move away from this narrative that plays too easily into a notion of racialized historical time. Furthermore, the African is definitely not the only "new" foreigner, nor does it explain what the grouping of the African exactly means if these groupings are made, as Cretton rightfully identifies, through discourses that rely on racist tropes, both biological and cultural.[72] If such is the case, then "African" as a name functions in this case less as a nationality or geographic location but as a racialized position (although it also does in that these *locations* are themselves racialized) and as such as that which *one* (i.e., *proper citizens*) has to move away from.[73] The category of the foreigner in Switzerland, hence, can show that Switzerland is invested in antiblackness through the foreigner being not only nonwhite but also close to blackness or Black.

Color-blindness and multiculturalism erase and, in the same turn, enforce the color line—the separation of the world into lesser and higher races. With origins in the United States' segregation in the Jim Crow era, the notion of the color line was later taken up by W. E. B. Du Bois and, in his theorizing, the color line became *the* question of the twentieth century.[74] But how are we to think the color line specifically in Switzerland, a country that is clearly color-blind, while also not having any data to prove the effects of the color line, as the very condition of color-blindness inhibits such possibility? This study thus proceeds by way of an engagement of the "lack of data" itself, of how there, where a lack is ascribed, lies a very loud answer to our questions exactly because a lack bespeaks that which identifies a lack, and/or silence relies on this remaining unthought and erased.

Historian Tyler Stovall faces a similar issue in his elaboration of the question of the color line in France, where there is also no data on race. He concludes that "the French experience thus suggests that the ultimate goal of strategies of racial exclusion may not be integration into a racial upper stratum but rather the consolidation of racial hierarchy through a denial of its existence."[75] Stovall comes to this conclusion by observing through his historical analysis that "the greater marginality and temporary presence of nonwhite labor in France brought about the triumph of a discourse of class over any conscious racial identity; unlike Americans, the French succeeded in so totally excluding people of a different race that the very concept of race disappeared from working-class life in metropolitan France, relegated back to the colonies where it belonged."[76] In Switzerland, there was neither this kind of historical situation, nor is Switzerland a country that had colonies. Nonetheless, there are a couple of ways in which this example of France allows us to think the color line in Switzerland. Firstly, the question of the grouping together of lower-class whites and nonwhites is a way to deny the existence of racial hierarchies (elaborated in Switzerland under the heading of the category denoted as "foreigners"). Consider also the fact that the color line acts hierarchical. Secondly, the color line is engaged in a national discourse of identity—that is to say, used to mark the identity of a nation, or space, in contradistinction to another (what would be the colonies in France).

Cretton elaborates on these two modalities in Switzerland. Firstly, there is a notion of Swiss national identity as raceless and white. Secondly, the problem of race is placed outside of Switzerland. Lastly, there is a grouping together of social classes that *covers* questions of the color line evinced by the category of "foreigners." Her opening paragraph speaks to the first two points:

> A common point of view in Switzerland is that there is no racism here. Switzerland is also well-known for being a neutral country, without formal imperial history and thus without "others" arriving from the colonies. Very often, the dominant culture binds those two ideas together, thereby producing this argument: There is no colonial population issue in the country (as there is in France, the UK, or Germany), thus, we cannot have racism in Switzerland.
>
> This way of thinking is shared among the majority of citizens within the country. It implies that since racism is due to the presence of indi-

viduals with an im/migrant background, the possible belief or mindset that race is a Swiss production is restricted. Such a system of thought reinforces and strengthens the idea that racism comes from outside the nation, instead of being produced within the inner society.[77]

We could read the first paragraph as something like an amplified version of Stovall's elaboration of France's dispelling of race to the colonies. Switzerland's involvement in colonialism is less obvious than France's, as they did not directly possess colonies. Nonetheless, as seminal research by Patricia Purtschert, Harald Fischer-Tiné, the organization Cooperaxion, and others have demonstrated, Switzerland still greatly benefited from colonial conquest and slavery through investments by both private individuals as well as cities or cantons in companies involved in the slave trade.[78] Additionally, the absence of colonial power is of course no argument for the absence of questions of race, except if race and racism are tied up in a limited and fallacious definition of directly holding slaves, or if racism is only where there are racialized subjects, such as, for example, the colonies.

Cretton points out discourses rooted in color-blindness and the assumption that Switzerland has transcended race or has always been raceless.[79] This belief requires that Switzerland is either multicultural—different "races" live together in equality—or color-blind—there are no races and no ethnic differences and everyone is simply the same, or at the very least the same within society as citizens. Cretton's case study in the west of Switzerland, the French-speaking part, provides yet another elaboration on how exactly color-blindness functions in Switzerland and underscores poignantly our elaboration of how Switzerland, by moving race elsewhere, is clearly marking itself as white. What makes Cretton's work particularly significant is that she analyzes whiteness through interviewing nonwhites *with* immigration background. One example of hers, which also involves border patrol, is of a Black Senegalese man who recounts often being assumed to be someone looking for asylum while crossing borders by train (i.e., the border patrol would order him to show his identification but not others).[80] Another case that Cretton presents is how other nonwhites attempt to distance themselves from blackness. Accounts from nonwhites, or immigrants, point to how they had integrated themselves better than someone else, someone Black: "On [the] one hand, when *negating* racism in front of the white, female, researcher, the participant is performing whiteness, which entails differentiating himself from 'others' (non-Swiss, non-white and lazy), aligning with Swiss natives, while at the same time performing efficiency (working hard). On the other hand, when producing racism, he differentiates and misaligns from 'others' (as in the case above, the dark-skinned Africans), meanwhile aligning with Swissness and the white majority group."[81] Blackness becomes marked as that which is undesirable, other, and not Swiss. In other words, whiteness takes its value in its contradistinction and separation from blackness. Lastly, we can also conclude that some people seem to be able to move away from blackness, as that which lies outside of proper national belonging, into whiteness/citizenship while others cannot, and yet others are always already Swiss (and not black).

Blood Relations

The management of the borders of citizenship moves along lines of blood. Similarly, the color line is delimited through blood, such as with the now-illegal one-drop rule in the US South, which was a cultural and legal practice that one drop of Black blood made one Black. They meet here, on the border of proper belonging, as a question of both biology and geography. To ask the question of whether the foreigner is allowed citizenship is to ask whether foreigners can shake their blackness, can leave blackness behind. It seems that Blackness is positioned as foreign whether from here (in terms of passport and/or physical habitation) or from somewhere else. Concludingly, the status of the foreigner is marked as that space that solidifies citizenship and its entanglement in blood and kinship (and racialization) as a site of contestation in the direction toward citizenship—the authority to control who gets to be a citizen and how sovereignty of the nation is claimed. The space under the heading of "the foreigner" bespeaks a kind of border zone, a kind of nonspace, that is a site of danger to here. Thus, critiquing antiblackness involves both making spaces welcoming to Black lives but also unraveling antiblack structures, because even when Black people are present, antiblackness can persist, since they can be ascribed to belonging somewhere else. Thus, the reduction of all Blacks as only visiting and being here temporarily supports a notion of a race-free society, one that is also not Black. As Didier Gondola argues in relation to France: "The preferential hospitality accorded black Americans in France served an officially sanctioned discourse proclaiming the absence of race discrimination and negrophobia" (or antiblackness).[82] Within these terms, any mistreatment of Blacks/Africans *here* is not connected to race but simply to immigration, an equally pernicious semantic construct used euphemistically in French official discourse in lieu of race. This kind of argument, or tactic, employed by antiblackness goes as such: there are Black people here, who are allowed to be here, or at the very least registrable here, then there cannot be any racism here, as Blacks are welcomed. But in the same turn, Blackness from here is erased, and all Black people here are marked as not really from here. In other words, if blackness can properly be placed as from somewhere else through acknowledging Black citizens located elsewhere, then this supports a view of citizenship, and presence, as not black.

Batumike in his contribution to thinking Black Switzerland, *Être Noir africain en Suisse*, elaborates on the difference between integration and assimilation in Switzerland and points out that to become Swiss one only has to integrate but not assimilate, or adapt.[83] I must mention here that if one wants to become a Swiss citizen, one has to *integrate*, be familiar with the Swiss way of life, and not pose a risk to Switzerland's security structures. Integration means "showing respect for public security and order; respecting the values enshrined in the Federal Constitution; being able to communicate in a national language in everyday situations, orally and in writing; participating in economic life or by acquiring an education; and encouraging and supporting the integration of one's wife or husband, registered partner or the minor children for whom one has parental responsibility." Furthermore, the cantons may have further criteria.[84] For Batumike, this is taken as a possibility gained—namely, that one can continue to foster communities with people from one's place of origin, exactly because one does not need to assimilate, one remains African.[85] At the same time, it is clear that integration keeps the differentiation between Swiss and foreigner intact, which is also to say, it keeps blackness and Black lives separatable from Switzerland. I read Batumike as providing a kind of resistance to an antiblack methodology, where our association with each other—for him, particularly, as African—as well as immigrants in general, allows for a kind of positive under the cover of this negative. This is because Blackness and Swissness cannot meet. But I also ask, How can we engage a discourse of Blackness *in* and *as* Switzerland that moves past the continued placing outside of Black lives and all things black from Switzerland itself? That is, in the continued claiming of Black lives as being tied to a continent, a time period, a mythical origin, whether that be Africa or slavery, the contemporary order of antiblackness can be erased. The continual discourse of Blackness from elsewhere can be a technique of antiblackness. It reveals also how an antiblack world uses a politics of recognition to erase antiblackness itself and to claim its own omnipotent authority.[86] But the Black Swiss context asks us to think differently about these kinds of questions. That is to say, that we may not reduce any instance of *Black* to a mere instance of an essentialized subjectivity. Rather, these terms open the very grammar of a world that claims to be all-encompassing.

In the underground of Batumike's positive spin, we may hear an unname, that of Black Swiss, not assimilation, not an erasure of Blackness or a relegation to an outside, but another kind of *here*, Swiss and/as Black. Poet, activist, and scholar May Ayim puts it beautifully with regards to Germany in her poem *"Grenzenlos und unverschämt: ein Gedicht gegen die Deutsche Scheinheit"* (Boundlessly brazen: A poem against the German fake unity) where she articulates a claim to being German, to being Black, both, and more, in excess of borders and imposed restrictions of belonging and existence, that we shall insist on being with our "brothers" and "sisters" by pushing at the "edges" of this hereness, of Germanness, of beingness, of belonging. And while moving, we refuse to be cut out, to be forgotten, erased, from here.[87]

Blackness is made invisible in Switzerland. As Katharina Oguntoye, May Opitz, Dagmar Schultz, and those Afro-Germans they interviewed point out in numerous ways in *Showing Our Colors*, in the case of Switzerland's neighbor Germany, Black life has been attempted to be eradicated from Germanness—not only intellectually and in discourses but also physically. Their earliest example is Wilhelm Amo, the Ghanian-born philosopher who received his doctorate at Universität Halle in Wittenberg, Germany, in 1734, where he later also lectured until his return to Ghana around 1743.[88] He left Europe because of the increased racial ideologies and because he couldn't endure the racism there any further.[89] Examples from the twentieth century include, from the Weimar Republic, the ostracizing of German white women who had "mulatto" children, which was very common due to the presence of African soldiers stationed with the French and English during World War I,[90] up to the extremes of sterilization practiced during World War II by the Nazis,[91] to simply not finding work or not receiving the proper documents because the situation of Black Germans during World War II wasn't thought about at the end of the war.[92] Hence, a plethora of forces at play since the early formations of nations in Europe have been pushing Black lives outside of Europe. It is within the social that Black Swiss are also made to leave—socialities are practiced as antiblack: "Basic terms of social human engagement are shaped by anti-Black logics so deeply embedded in various normativities that they resist intelligibility as modes of thought."[93] That is to say that while social interactions invested in antiblackness as a function of affirming civil belonging, in the form of proper citizens, "continually produces Black people as out-of-place," blackness bespeaks sociality in excess of measures of sovereignty (as normativity's entanglement with properness as a measure of proper human life).[94] Blackness is utter sociality as it is experienced and practiced with others, even when such others are not present—a togetherness in excess of togetherness as self-contained in one's meeting. And it is in its pouring over the frames of the white gaze, that structuring of civil society on antiblackness, that blackness is hyper-social; it is where social and private converge.

Saidiya Hartman, toward the end of *Lose Your Mother*, hints at the way in which blackness practices an otherwise sociality beyond the measures of the state: "It was a moment of fleeting intimacy. This give-and-take was not a matter of blood or kinship, but of affiliation. We were the progeny of slaves. We were the children of commoners"—a commonness on the basis of no blood.[95] What is the relation between the social and the political or the sovereign state? What does Black life do to citizenship? Is it conceivable that relations can be rethought through a matrix of the poetics of blackness, of blood in excess of measurements? The practice of Black studies, as talking about blackness and Black lives, speaks of and asks us to think of socialities in excess of a proper, and a prefixed, notion of what relations can exist.

The border of Swissness is articulated in a myriad of spheres. The fact that race, antiblackness, and blackness are inarticulable in Switzerland—erased—reveals that race is structuring Switzerland's self-image. Antiblack methods of border control hold Black lives in the border zone through asking for identification and in this always already having identified them as a (possible) danger, as uncertainties, as unknowns. The ask for identification seems to function as an erasure of the identification of those who aren't asked. It seems that such an "ask," or demand, is nothing other than the attempt to locate all such uncertainty in Black, which means erasing uncertainty from those who properly belong. It makes the inside both absolutely one thing and at the same time erases this one thing. It is the articulation of race as well as its erasure. It is the articulation of *a* normativity in multiple domains, including phenotypical traits (i.e., racialization) exactly because it happens on the basis of such. This normativity inside is articulated through the identification of the other as the erasure of the identification of the inside. The very asking for identification is always not an ask to a citizen; it is always a demand to a possible noncitizen. This racializes the nation, makes it have a normative phenotypical appearance, exactly by erasing all questions about its appearance and placing all questions of appearance onto the other.

In their resistance to showing identification, Black lives are doubling down on their unidentifiableness, shifting what identifying may mean, moving past a space of controlled unidentifiableness as opposite for the claim over proper identification. The ask for identification implies that they are unidentifiable, and it is the unidentifiable that is absolutely marked.[96] Black lives protesting are uncovering the normalized inside—that is, the world of proper identification built on antiblackness. Their refusal both identifies antiblackness, as well as Black lives, beyond authorized recognition, for it shifts the very meaning of the unidentifiable and identifiable. In other words, by refusing to answer the question, Black lives are opening the closed world of identifiable persons (who are always identified against the unidentifiable marked as black). Black lives are teaching us to ask questions about the world by opening the ask. The question is: Who is us?

3

Interstitial Listenings II
Blurring the Hold

Blurred Visions

Perspectives and Blurred Colors is the title of a piece by Jérémie Jolo that hints at a kind of shift in vision that we must engage to listen beyond national belonging as a question of a delimitation of proper lives, unmarked good citizens, or authorized kinship. Written in 2022 as a way to "digest" the war in Ukraine, it bespeaks a kind of alternate sight that falls over from a fixed gaze into care for each other and the world.[1] Blackness teaches us to hear each other because it puts a mirror to the ask of identification. If the asking for identification is the making uncolored of the inside of Swiss citizenship, then it is the protests and refusals by Black lives in Switzerland to show identification that uncovers Switzerland's colors—including its antiblackness, but also much more than that. By blurring colors and showing perspectives, Jolo is asking us to think about what the sound of lives in excess of proper belonging as unmarked proper citizens might sound like. In *Perspectives and Blurred Colors*, we hear a legion of voices, all Jolo playing the clarinet as well as singing, recorded and stacked in a looper. These voices move together in a seemingly rigid formation, in the same rhythm and in consort. What is striking, though, is that despite such march-like quality, when listening more attentively, one starts hearing all these variations within the singular lines, variations of rhythm, pitch, timbre, and more. Here, this initial semblance of fixedness, this kind of hold that is the groove, is revealed to be insufficient to account for all those voices that apparently make it up. By listening in this manner, we hear the voices exceeding any one fixed gazing, any kind of reduction to a grid, a rigid sameness, of groove and rhythm.[2] I hear this as a kind of opacity, like how the complexity of lives is the opacity of the world as the continual failure to hold on to one view; like the black opal, that stone that always pops into my head when I hear this word *opacity*, a stone with so many colors blurring into each other through the clarity of reflection of invisible blackness. Opacity bespeaks the complexity of lives lived in relations that, as French Caribbean philosopher Édouard Glissant reminded us, are a fundamental right—how we must remember that we can never be fully known by anyone (not even ourselves).[3]

Jolo, whose father is from the French Caribbean island of Guadeloupe and his mother from Switzerland, was born and raised in Biel/Bienne, Switzerland, and studied music from an early age. A clarinetist playing the classical canon with ease, he also creates his own works, primarily using a looping station playing, singing, and layering his voices over each other. He is astutely aware of antiblackness in Switzerland, as becomes clear from his interview published in the *Bieler Tagblatt* in January 2022. At the same time, he points out how in music there is something that escapes this antiblackness. Music as a field, whether practiced by amateurs or professionals, is, of course, itself still entrenched in antiblackness—for example, through demarcating proper belonging. But as Jolo points out, when music is made, there is something that cannot be spoken of within antiblack paradigms.

> I know that brass bands are something originally Swiss, very traditional. So I was internally prepared for something, when I started there [Bellmund and Sutz-Lattrigen music societies]. . . .
>
> However, the musicians met me with openness from the start, which was probably also due to the fact that some of them already knew me. And the music did the rest. It's something that connects people anyway—perhaps because you listen more than you look. At musical competitions, however, I have already overheard comments from other music groups such as "hey, hey, this is our own festival!," in other words our Swiss festival.[4]

Uncovering some ways in which antiblackness shows itself in the musical field in Switzerland, in very explicit ways, Jolo's remarks move us beyond such scenes. I hear his statement, that "you listen more than you look," which does not exclude looking but emphasizes listening, as a hint to these blurred perspectives that blackness bespeaks. Black lives tell of a togetherness in excess of any account of sovereign domains and against the disidentification from Black for a colorless, antiblack world. Here it is obvious that identification happens so that they—those who call him out—can claim their own belonging as proper, as always already given, through questioning Jolo's. But what might he mean with: "The music did the rest"? I can imagine it on the order of a question of having to learn how to listen. They all must listen to each other and reveal their sound; that is how they listen. They all must acknowledge their opacity to each other. No one has claim to remain transparent against an other that is opaque. That is, no proper citizen, who is of the order of transparency as being inside the known, can be claimed on the opposition to an opacity of those who are Black. It only makes sense when we hear it with Jolo, with the antiblackness against Jolo. In their following Jolo, as musicians whom he conducts, they must give up their claim to knowing how to listen. They must give it up to him who helps them hear each other, so they may become an ensemble. They must listen to Black lives.

Thus, I don't hear him suggesting to ignore antiblackness but rather that the very critique of antiblackness always also means that there are Black lives. And in a second step, he turns our attention to a practice that thinking with Black lives proposes. It is a protesting and exceeding of the confines of an antiblack world that delimits who belongs based on a prefixed definition of life. Let us here simply state some things that Jolo and his works announce to us, as a kind of Black radical thought or practice, and I will elaborate further on these aspects throughout this book.[5] Blackness is livedness, is the music of a sociality that exceeds the white gaze, the antiblackness of a frame that claims originality of kin measured in sounds, appearances, or performances. To be with blackness is to be together in music, together in sound, in excess of being together, because togetherness might imply static, singular beings that meet. Blackness is that which speaks of a together-apartness that is before any individual and which forms them all—it is radically before any togetherness in contradistinction to apartness. Or we might also say: it is the way in which apartness implies and is with togetherness and vice versa.

Karen Barad brings us the term "together-apart" in relation to quantum physics and their brilliant reading of Niels Bohr as a question of the cut, which we might think of as a crossing of borders: "Cuts cut 'things' together and apart. Cuts are not enacted from the outside, nor are they ever enacted once and for all."[6] This idea comes from Barad's engagement of the famous double-slit experiment, which proves that the fundamental stuff that makes up our world is sometimes behaving like particles and sometimes like waves. And it is the blurriness of reading waves or/nor particles that this notion of together-apartness bespeaks. In other words, rather than saying it is particle or wave, it is rather that it is sometimes a little more wave and a little more particle another time, but still always *both*. Instead of conceiving the wave and the particle as two different states, with a choice to be made between one and the other, Barad points out: "Instead, Bohr insists that what is at issue are the very possibilities for definition of the concepts and the determinateness of the properties and boundaries of the 'object,' which depend on the specific nature of the experimental arrangement."[7] That means that whether it behaves a little more like a wave or a little more like a particle, is not a question of that which is measured, but of both the measured as well as the things involved in measuring. The whole setup, the equipment, the tools, and even the observer are all part of why it appears sometimes more as a wave and sometimes more as a particle. An entanglement between states of matter and observation, as well as experimental tools.

In Jolo's work RED, 2018, this blurriness between particle and wave is centered. As he writes in the program notes, the music, which he also calls a "*Frequenz-Chaos*" (frequency-chaos), is made from musicians who have their own sounds as well as from the sound of the whole group.[8] The piece has a groove reminiscent of Moondog's repeating modal chords and short melodic lines. He combines these pulsating harmonies with a series of filters that move us into different kinds of reverbs or sonic spaces. A myriad of voices, created through Jolo's singing and clarinet playing, again looped and layered on top of each other, dance around each other and relate in some kind of opaque but poetic way. There are these delays of rhythms, chained, like some kind of blurriness between particle and wave, where we can lose ourselves in listening, where it is unclear to us where we are positioned in relation to the beat: the observer's view becomes fuzzy and the possibility of holding the listened to, the observed object, in a fixed position, fails. In some ways it reminds me of Julius Eastman's music, like those piano pieces played by Swiss piano quartet Kukuruz, where individual musical pitches become fuzzy and the note's singularity is revealed as an ambiguity of what its place is (like a black hole).[9]

Jolo puts it this way: "One could jokingly tell somebody to listen to the individual voices and the effort I've put into each one when recording them in

2018! Oh, how impressive. And now listen to the big picture and the whole resultant sound, and pay attention to all the repetitions! Oh, how boring. And now close your eyes, go into a meditative state, and it is, through a certain blurriness in perception, music after all."[10] I hear in this statement about what the music does and how it is put together a recognition of this more than in-betweenness, the ways in which the very duality between movement and object, momentum and location, is not where the music happens. Listening becomes problematized: does one listen after a cut, in a mode of duality with waves and particles, or so-called before, which is also within this break, as a blurriness (an underground from which duality is holding itself off from)? It is not so much about a more fundamental shared quality or overarching wholeness, as Jolo mentions, that is not the music, but rather about a sitting with an open ear in the moment of their articulation, in an interstitial kind of way.

Frequencies of individual musicians don't express the music just as the composite sound of the band doesn't express the music, but rather the music sits in the opaque relationship not-in-between them.[11] Like a quantum field, both field and points (waves and particles) make up the field. The field is not a self-existent container of particles or waves, nor *are* singular particles and waves the field. Furthermore, the relationship among those sounds is opaque, is never fully knowable, and it is in this unknownness that the field, or relations, come to play from and as. Thus, this is music. Jolo concludes with "*Es lebe das Chaos*," which can be translated as "long lives the chaos," but a transliteral translation would be more something like, "It lives, the chaos."[12] This bespeaks another kind of way of relating, not because chaos is the opposite of order but because this kind of order, the one of chaos, is the aliveness of ordered things, of things that bring forth ordered sounds as music. Maybe one could hear Jolo's statement in conversation with Glissant again, with how, out of opaque relations, that are musical or poetic, in Glissant's words, comes to the fore a "chaos-world," a world in opaque relations.[13] And so it is clear that what we speak of here is something born as thought, in the belly of the slave ship, in Blackness. This kind of story of relations is one that speaks of a togetherness and aliveness that flows across holds and does away with being in opposition to what is unknown and unforeseeable. It bespeaks the making of relations across what is proper, the here as in relation with there, and not a violent opposition to the there for a claim to a proper here, a claim to land against *other* lives. Glissant calls it "poetics of relation." The poetic with relation is what shifts relation and poetics out of the known into the unknown: poetic relations rely on the interconnectedness between different particles as well as their opacity. Relationality is the root and at the same time in excess of (or not) itself—poetic relations are relations in overflow of relations.

Transmutation of Listening

Slowly it becomes clear what transmutation Blackness is performing within and in excess of the hold, the cauldron, on both its substances and the hold. Musical objects, and musical form, cease to be absolutely distinguishable in these two pieces. And such is not so as to articulate a kind of irrelevance of their appearance, but rather to point to the very way in which they appear. This means to bespeak the listener or observer and their role; it means to think about the space between things, that border between one and the other. By bringing out the many shades of this interstitial space, Jolo teaches us that this is the continual falling into the music that makes the music in the first place. It is like when disidentification ceases to be possible, that closure that fashions a controlled border because we never allow for the closure of this interstitial space. We keep associating across boundaries of relations—that is, what makes us opaque—because we never claim ownership over all knowledge this is what brings us together, as listening to each other's stories, and this is what allows us to live, as hearing our actions as of and for the music, that unknown chaos-world.

We can think this as a question of unanswering, where the question is asked continually without a possibility of its end.[14] It is where questions become our continual sitting with and in the interstices of things. Unasking is a nonmove to a nonspace from which such ask comes from, from where asking and not asking—in forms of answers—come out of and return to. A certain sitting with the question's appearance and disappearance that is concerned with what appearance carries as its underground. Or, to truly complete this transmutation, where the very asking of a question was never about its closure in an answer, but answers become only another question, another listening from, to, and for the unthought.

The reopening of the question as the refusal to hear an answer was also W. E. B. Du Bois's incipit to thinking about Black lives in the United States—the Negro problem as the question of the problems of our world. In this scene, which commenced with Du Bois's *The Philadelphia Negro*, the first ever undertaken sociological study of Black lives in the United States, published in 1899, the study of Black lives opens the very study of the world and lives in excess of what is taken as given. While the notion of the Negro problem was not invented by Du Bois (the notion emerged to thematize the problems of the new Negro citizens after the passing of the Fourteenth Amendment in 1868), it was his life's work that shifted it: from a question of the Negro as a kind of problem or as the location of problems to the very study of lives and the world that exceeds any problem as negativity, and even as reflection of the very conditions of the world in a certain moment in time or a place in space. Du Bois brings us to study the world through the study of Black lives.[15]

Du Bois's project as political force moves by way of the unasking of a question posed as already answered. It is under the heading of cases that we uncover antiblackness in Switzerland. It is under the heading of cases that we start hearing Black lives in Switzerland, and it is when the case becomes example that moves as an opener for thought rather than as exception that we start hearing their questions to/for us. And so we follow Du Bois, who announced the study of Blackness as the study of the world. The example itself, the case, is before that which is *an* example of (or *negative* exception of). It is a kind of opener to think, to remake, the world. It is the very opening of study and thought, as an unasking, as the opening of questions and thought, that keeps us in this interstitial listening.

In "My Evolving Program for Negro Freedom," Du Bois reflects on his life and practice, and he ends on a warranted pessimistic note about the future of mankind. In a brief elaboration, he stakes out his own practice as a movement of a couple of errors of judgment, followed by a commitment to (continue) *doing the work*: from the assumption that truth will be accepted by whites, to the assumption that white America would want to defend democracy above all else, to his then-current project, where he admits that because everyone is situated in customs, in both practices and thought, that the "race problem" will increase. That is to say, to paraphrase him, that "the mighty of the world" will not give up their place easy, even if this giving up achieves the emancipation of mankind and the realization of democracy.[16]

Du Bois's observations are poignant, not merely as personal autobiographical facts but also as a general situation of what study, as black study, entails. It is within his enumeration of a process that we can find something that *works*—not only to learn about some kind of fixed world but as a way to change the world through study, where study becomes the very changing of worlds. It bespeaks what Du Bois's work did and does to change the world. It is the way in which this three-part process reflects a kind of formula, a kind of magical mathematics, something like an *Immeasurable Equation*, like those blueprints written by Sun Ra. It can be translated into how there is, in a first step, the refusal to show identification, which is to say, the speaking out against antiblackness. In a second step, then, is revealed the way that antiblackness is entangled in everyone's positioning in the world because it is the structuring paradigm of the world. We might also call this the conditions of and as the world today. And the third step exceeds this very stepwise progression, because it seems that this third step is first, second, last, and before them all. It is this last step, the call to continual study, that must be heard in excess of the very tasks at hand, because it is beyond them, it is in excess of even the conditions that call it forth (but it is there in their underground). We might say it is this opening that unasking bespeaks. This is the way in which the refusal to show identification bespeaks the refiguring of the world through making Black lives matter.

At the beginning of this essay, where he outlines these three steps, there is actually a hint to how Du Bois attempted to practice this, presented in the form of a metaphor, of an example. It is revealed in his distinction between the social register and the city directory. Du Bois explains:

> The hope of civilization lies not in exclusion, but in inclusion of all human elements; we find the richness of humanity not in the Social Register, but in the City Directory; not in great aristocracies, chosen people and superior races, but in the throngs of disinherited and underfed men. Not the lifting of the lowly, but the unchaining of the unawakened mighty, will reveal the possibilities of genius, gift and miracle, in mountainous treasure-trove, which hitherto civilization has scarcely touched; and yet boasted blatantly and even glorified in its poverty.[17]

The social register in the United States lists the most affluent members of American society. But not everyone who is rich can get into the social register; membership is mostly established through kinship, with the rare occasion of including new members through sponsorship by someone who is already on the list.[18] The city directory, on the other hand, was, in the United States in particular, a listing of people who stayed in a city and included how long, and might also include their family members, address, professions, and sometimes ethnicity.[19] Additionally, city directories were published yearly and so it reflects also who moved where or who lived in a space from when until when. (The social register was also published yearly.) Du Bois's metaphor, thus, points out a differentiation between belonging enforced on the order of wealth and proper kinship and a more continual engagement of lives who live at a place. There is also, of course, the possibility of turning the idea of the city directory into an enforcement of proper belonging through the use, or addition of, other methods of control. For example, it could be turned into measurements of migration and again used in the service of determining proper belonging.[20] But at the same time, the existence of these kinds of documents can prove very troubling to proper national narratives. Such is exemplified, for example, in the case of the *Book of Negroes*, a book listing three thousand British loyalists and freed Africans from 1783 who were relocated to Nova Scotia, Canada. This work proves the presence of Black lives in Canada and allows for the problematization of antiblackness in Canada and the reconstruction of Black Canadian history.[21]

This difference between two modes of listening to lives introduces a problematic that Du Bois elaborates upon in his 1905 essay "Sociology Hesitant," a manuscript that was only published in 2000.[22] In this essay centering the practice of sociological research, Du Bois points out how there is, regardless of whether it is within an erroneous attempt to mark the categories of social studies as wholes and parts, as pregiven or as absolutely unmeasurable, an underlying shared assumption. They all imply that there is a set of measurable laws, as sociology, that are put in contradistinction to an unknowable space under the heading of life (as that which sociology studies).[23] After uncovering this correlation, he elaborates on how sociology should "accept the realm of chance" and practice science that works with it, rather than have a series of absolute laws in contradistinction to an unknown realm of life.[24] That is to say, that rather than reifying sociology as a domain of knowledge, with prefixed categories of analysis and a separate space of the unknown, this chaos of *real* life, Du Bois proposes that it is the *work* of *studying lives* that sociology should practice. Thus, the study of lives shall continually change the very categories and laws that are then, in turn, used to study life. Like the city directory, sociological research is never complete, and its annals must be rewritten over and over again through the engagement of lives and their movement, their relations and the world. This means knowledge is not about the production of some kind of space ridden of nonknowledge or the unknown. Rather, knowledge is the unknown creeping in. It is about a shift in the relationship to the borders of knowledge, where they cease to be borders as something that denotes control and become interstitial listening. It is about how there is a crossing of borders before they are authenticated as borders, how the crossing of borders reveals how borders are supposed to be crossed—not only on the order of what they are in service of but rather in excess of that inside's control, which means it is the changing of what is inside.

This is about how to listen to and *for* Black Lives, also in Switzerland: How can we speak of something that both escapes law (or population measurements) and chance in their formation in contradistinction? If Blackness is real in Switzerland, then it is not so as a function of registering, as there is no statistical data collected in Switzerland that could account for them. The available measurements to researchers in the form of immigration background are not sufficient as a measure of Black Swissness—within the reading of data of migrations is also a certain question of incalculability of Blackness, Blackness as located in Switzerland, because it effectually only measures what falls outside of Switzerland as Black. Nor is it possible to account for Black Swiss through a space of the unknown as outside of measurements: as has been evinced, the very unmarking of the citizen is the relegating of Black lives to the border. Thus, the current (not)measuring is part of a color-blind modality that erases Blackness in Switzerland and thus this unmeasurable space becomes overwritten with what is measured.

The question of listening for Blackness in Switzerland is through both measurement and nonmeasurement erased from within, from life here—it is through not listening that all things race, racism, as well as Black are placed to the borders of Switzerland, and its inside becomes unmarked. This is an antiblack gaze on a national level, where Blackness becomes abject and absent from whoever is gazing, as holding in contradistinction. Thus, this gaze is also a modality of studying society that does not really study, so as to keep what *is* from what is unknown, what is not. This happens through following categories and laws that are prefixed and measured in contradistinction to a space of the unknown—marked by and as Black. By pointing to the city directory, Du Bois is marking a space of sociological research (or we might also say, listening) that flows with and as the uncertainty of lives. This interstitial listening sits at the border of known and unknown and relates them, over and over again, against any enforcement of this border and the proper movement—relations—across it. In the underground of borders, that delimit categories so as to protect an inside that can, as a result, never be changed, only be expanded, lies this interstitial space. This interstitial space is how we may speak of an excess of the control claimed in the border zone, an underground of borders and their crossing—the very crossing of borders that comes before them. That is Black studies, this poetics of relations, this chaosworld, this music from Black Switzerland, that we hear by blurring our views beyond what we know. As Swiss rapper Thierry Gnahoré, a.k.a. Nativ, says, "*Hüt isch ä guetä Tag für ne Change*" (Today is a good day for a change).[25]

Gnahoré grew up in Niederscherli and Bern and has been living in Biel since 2017.[26] Nativ is a prime example of how Black Swiss associate with Black lives elsewhere, such as in the United States. Here, referencing Barack Obama, Nativ's music was also present in the Black Lives Matter protests in

Switzerland.[27] While popular Black people and cultural artifacts, mainly from the United States, are often consumed by audiences across Europe and can be a way to erase Black Swiss lives here, there is another story told when we listen to Black Swiss lives. This erasure and consumption is often accompanied by the concept of "authenticity"—that is, some kind of authentic blackness that comes from elsewhere than Europe or Switzerland.[28] This space of proper blackness is delimited, fabricated, and circumscribed by those who are not black.[29] That is to say, the claim to defining what black properly looks, sounds, and moves like is equivalent to the scene of policing: blackness is absolutely defined as certain things and from certain places. This is why Cikuru Batumike looks critically at the claim to being inclusive by welcoming Black lives from abroad at international jazz (and pop, rock, etc.) festivals: "Time to forget those Blacks who are neither in the limelight of applause nor passing through; Blacks who live, themselves, on a daily basis, in Swiss realities."[30] But Nativ and his music demonstrate how Black lives use and refuse the ways in which the media give access to Black lives.[31] That is also to say that an engagement with the music of Black lives without confronting the antiblack structures of the world can mean the continuation of those same structures. The confinement of Black lives, and blackness, to a few sounds or genres is the way in which the extraction of vitality from blackness toward an antiblack world can be facilitated. To fight antiblackness thus means to resist and refuse the claim that Black can be delimited by any appearance or already made container by an antiblack world. In a world where all official and commonly known things profess one's own nonexistence, such as the Black Swiss experience, we must always listen in excess of categories and available networks. What does it mean when change even exceeds what is commonly understood by change? That is the practice Nativ asks us to engage in.

4

Afrofuturist Archeology
Citizenship and the Delimitation of Life with Death

Identifying Life

Antiblackness's claim over black and its definition is continually put to use to delimit what is and who can be in a space—who may be a citizen. Since the very beginnings of the definition of modern citizenship in the nineteenth century, there were aesthetic performances of antiblackness towards the definition of national belonging. National exhibitions demonstrated what living in a nation *looks* like, but also how *others*, and specifically Blacks, supposedly live like. For the 1887 National Exhibition in Switzerland, a fake "Black village" was created, full of stereotyped African ways of living, with two hundred Senegalese people on exhibit.[1] Similar to musical performances, the performance of antiblackness stereotypes, delimits, and claims ownership over black people to claim omnipotence and authority over space and life.

In this chapter, I unpack the relation between the formation of sovereignty and nationalism rooted in the nineteenth century and antiblackness. Demanding identification of Black lives is the aim to control blackness and the formation of controllable borders. This kind of identificatory practices are performed not only by state officials but individual citizens as well: such as through touching Black hair,[2] through children's games,[3] through erasing Black lives and their practices in a space such as Black history, through asking, "Where are you (*really*) from?," and other forms of identification (and delimiting) of Blackness—as ways to mark one's own proper belonging.[4] In other words, the scene of identification as a questioning of Blackness's presence is performed not only by police and border patrols but also by individuals of a society. But such claiming of power happens through the distancing of the underground of these borders as the claim to having authority over the uncertainty of belonging. In other words, the border's underground, that which it is supposed to keep in control, the possibility of uncertainty, of an outside, is dispelled from the inside, from the nation and its citizens, through claiming absolute control over borders and being able to hold them at a distance. These kinds of borders, this kind of border policing, are there to dispel the uncertainty of the unknown outside, of blackness as something that in this equation becomes the very changing of lives, the very possibility to be in touch with the world and each other. Black Lives Matter protests became so powerfully disruptive because they were and are enmeshed in a network that forms itself and reforms, and informs, itself through this coming in touch with each other across and within a continual crossing of borders, across life and death, across what can be told and what can't.

Let me take you on a journey into this interstitial space, and we'll travel through and with the work of Frantz Fanon. The seminal Martiniquan scholar begins the fifth chapter of his groundbreaking work *Black Skin, White Masks* with two phrases directed at him by a little boy during his visit to mainland France: "Dirty Nigger!" and "Look, a Negro!"[5] This scene of *identification* leads Fanon to his elaboration of the "white gaze" as infantilizing and enclosing—enclosing through the skin in particular.[6] In this elaboration, Fanon thus centers a negative, a debasing, view toward blackness found within both the "Negro" and "whites," but in the Negro experienced as against themselves, and as absolute fungibility to the white world. He elaborates that blackness emerges as the other and does so in relation to "the white man."[7] In this, Fanon thus thinks blackness as emerging in opposition to whiteness—as whiteness's other. His explanation takes the form of a narrative: Fanon states that before arriving in France, living on his home island of Martinique, there the Black man is not confined to being "nègre."[8] In other words, the reason why being black is in relation to the white man is because only then is black absolutely locatable, unrefutably, in one person so that the other could never be othered by this person (or be *touched* by blackness). But what does this temporality that Fanon invokes in his narrative explanation indicate? What is this life before France, in Martinique, about?

I want to think about this question of national belonging as expressed through a gaze that localizes blackness outside in a more general sense. It seems that this kind of gaze, and a sense of proper belonging, belonging to a nation and a civil society, are one and the same. Fanon elaborates that Blacks can act "black" in Martinique without *being* "black" while among fellow Blacks, but here the definition of "black" is still one of the white gaze and thus the white gaze is already an absent structuring principle.[9] One way of reading Fanon here is that blackness is fabricated. But this causes a kind of surplus; namely: How can the Black man have no power over the white gaze while also being outside of it in this previous space of the island in the Caribbean—that is, before visiting France—while it is still possible to disengage the negativity of it (evinced in being able to stop acting "black")? We might say that in revealing that black is nothing other than the negative other of the white gaze, Fanon points us to the way in which black must exceed this definition (the one it must assume in this scene of gazing).[10] There is a question of what this confrontation with the white man that Fanon relays to his readers does—not only in the sense of the individual's personal sensation of status as separate from civil structures but also as a certain revealing of a second-sightedness of the Black man that enters the picture when the claim to citizenship, as this gaze, becomes troubled.[11] The second-sightedness is the way in which the author of the text, Fanon, is able to recount the white gaze without being able to hold it, thus falling out of it and seeing it in this very process, and being able to study it and its effects. In this second-sightedness—that is, the white gaze's impossibility in/of the Negro—there is a critique of a certain modus operandi of the white gaze that was forgotten while also present, in some registers, in Martinique.[12] We can hear it again in two ways. One way is that there is also a normalization of behavior structured by a white gaze, bespeaking the globality of antiblack structuring paradigms. In the second way, there is also a possibility to hear another underground—namely, *What if it is about an impossibility of the white gaze, expressed exactly in the ability for Blacks to act black?* If, indeed, the black man is black in relation to the white man, then there are two ways of thinking this scene: either acting black means that Blacks can become white through locating the acting of black in another person. Or, alternatively, acting black is also acting whiteness, or the white gaze, because this acting black is acting out the white gaze's view.[13] But, since black is only in relation to the white man, this acting of the white gaze (through acting black) can here never acquire a kind of realness, in turn separating a real white gaze and an imitated one, and such happens always through a debasing view of/against black. In other words, since Blacks are black in relation to an antiblack world and have no power to turn this around, their acting is always merely an acting, an imitation of a proper white gaze. Thus, we might have to draw two conclusions. One, the white gaze bespeaks a structuring paradigm of (national and societal) belonging that is global. And two, Black lives is where the white gaze always breaks down because of the impossibility to properly be white, except as a momentary and only partial act against "his fellows"[14] since, in relation to the white man, the Black man will remain Black.[15] Thus, the white gaze bespeaks on the one hand a societal (and psychological) performativity (which I will take up again in chapter 10) and, on the other hand, it suggests always also a global material, symbolic, and structural paradigm that, because of its claim to omnipotence, cannot be evaded. But, at the same time, there something happens, in this acting of the white gaze, that is both a protest through and a celebration of the second-sightedness of Blackness—an excess of being in opposition to an *other*.

Revealed is a bizarre relation between color-blindness and the white gaze that breaks down in Blackness's refusal of the white gaze, which is also a revealing of a certain articulation of the white gaze as a global structuring principle in civil society. Thus, might Fanon also be pointing out how the very structuring paradigm of modern national belonging is continually engaging a "white gaze," a gazing that holds Black life, and others who are close to what blackness symbolizes, in opposition, at the border of belonging, life, and citizenship? A hold that provides the very foundation for a claim of authority over who gets to be a part of the world, of society, of proper human life? This is why I call the white gaze an antiblack gaze because it uncovers how there is a claiming of being of stature within society, or a nation, that happens against blackness, an outside exemplified and personified in Blacks, that must be renounced to properly belong. And this renouncing is also the very upholding of a border, a limit, of what it means to exist, to be oneself, and also the forming of an absolute end, which is the controlling of the outside of what is allowed to be here.

Fanon bespeaks a certain impossibility of blackness to escape relations that can be read in two ways. Firstly, Fanon elaborates that the Black man is black only in relation to the white man, thus to the white gaze with which the white man possesses the Black, and there is no reversal of such gaze from Black to white.[16] Secondly, there is thus also expressed a certain disappearance of the blackness of the Black man when there is no relation to the white man. It seems that this means that blackness itself, and not just the "man" that *carries* it, is outside of relations of power. The blackness of the Black man cannot enter into relations, as it has no standing to make relations. This bespeaks a liquidity of the Black that is a certain nonlocalizability of blackness in a world where everything is simply decided by its position in relation to this white gaze. If the identification of the Black is always only the white gaze's power, then this identification of black is absolutely not the Black's. It becomes *a property held by the white gaze*. But such cannot be held by it, because it mustn't be black, not even close to it. It must erase it from here; thus it has to be held in its absence, outside of it. This outside is the outside of the subject—the subject being a citizen or a nation and the border is a no-man's-land. This liquidity as a question of nonrelation asks us also to think "black" as something like a commodity, as something that is like capital liquefied.[17] At the same time, this liquification of the Black is a solidification of the position of the white gaze, of those that can own, or possess, blackness outside of them. Ownership and possession are thus not (properly) accessible to the blackness of Blacks.[18] So the question of belonging and possessing citizenship, or the ability to be a citizen, is somehow entangled in an antiblack gazing, a registering of blackness as always already outside, on the infinitely small border zone held off by those who can have a standing in a world of relations for economic exchange.

Blackness has, as Fanon elaborates, no resistance to the white gaze. What is that without resistance, that which is not a resistor and is thus beyond measurements of capacity?[19] It must be concluded that the white gaze does have resistance—that it is the measure of resistance itself. That is to say that what seems to be behind Fanon's articulation is also something about blackness that escapes Fanon's elaboration because such elaboration follows measurements possible only with resistance, as an elaboration of antiblackness.[20] And it is here that the liquidity of blackness marks something that escapes the confines of what is hailed in "Look, a Negro!" There is in this hailing, where the white gaze marks itself and those who can practice it without fail as proper citizens, also always a space of blackness, which, through not being measurable/resistant, is beyond the hail's instantiation of relational contracts between men, or white against black.[21] The blackness of the Black man is how Blacks exceed their own positioning in the world as the negative other because they must be the very absence of the presence of those that can gaze: proper man.

In some sense, the scene of antiblackness that Fanon's case presents must foreclose blackness from itself so as to be able to hold blackness as possession, as exchange good. That is to say, that in this scene where the white man is not available to the Black man, the Black man seems to also flow beyond the white gaze as historically, and more, contextualized by antiblackness.[22] This context is one of politics and a certain notion of life that are both made in the delimitation of life, into political life and some kind of lesser life, for the establishment of citizenship or civilized life. The white gaze is tied to national spaces and is thus also in relation to citizens or the articulation of citizenship. Thus, the question of blackness and citizenship, or nationality, is also one of the delimitation of life as announced by the white gaze, by this antiblack registering that fixes relations through the exteriorizing of blackness. How is Black life in particular outside of citizenship, or as *problem* within it, while also escaping the very world of citizens and nations? If this gaze holds blackness off, not only from the "white man" but also from the "Black man" because the Black man cannot ever be Black, as blackness is already owned by those gazing, then to steal this stolen blackness (which is inevitable for the Black) is to undo an antiblack conception of life that controls who gets to be alive here, in this space and in this time.

Afrofuturist Dig

I want to flip the script, turn the whole world upside down.[23] What *is* shall be seen only through the eyes of what is not. What if we think alongside that which is outside of relations, outside of resistance, and announces another way of living rather than in opposition to others? Can we follow this blackness instead of this world in relations where *black* only denotes its end, its outside, its notness? I want to think after the end of the world, like the Afrofuturists who announce to us, from a spaceship that is an arkestra (ark + orchestra), and flies with the sun: "It's after the end of the World. Don't you know that yet?!" The phenomenal musicians of the Sun Ra Arkestra repeat the phrase: "It's after the end of the World. Don't you know that yet?!" while floating in outer space in a spaceship (the mothership?) at the beginning of the film *Space Is the Place*.[24] What if space is this outside, this endless nonspace, where all spaces, these places we might call planets, come as the sound of this vast nothingness? Like when the Arkestra starts with a giant sound with all the possible notes and timbres in it just to tune us, the audience, into their music, to begin their concert, to open space itself.[25]

Thinking with the outside means thinking about that which is claimed by the inside as its other, as that which the inside does not want to be, as that which the inside calls absolutely different, an error, an exception, something that should not be. Maybe Fanon pointed us to an object, an unidentifiable object that escapes grasping, only because the white gaze, or those on the planet, those who try to hold this thing, have a mistaken view. What if they can't see the mothership because in order to see it they need to come to grips with their own position in an endless outer space? Let me say it like this: this violence of identification is really about dis-identification, which is the delimitation of (as into) an outside as the attempt to erase all uncertainty from the inside.[26] Misidentifying something unidentifiable puts one in control simply because by making it unidentifiable, those who do so become the masters of their world—they get to decide what is. Like the white gaze as a way to identify black as not here, as over there, somewhere at a safe distance from *us*, the proper citizens.

The figure of the exception of citizenship, of citizens, and thus also of nations, has been theorized under the heading of *homo sacer*. It is a term *dug* from ancient Rome, where homo sacer was someone who broke the law and as punishment was expelled from society and could be killed but not sacrificed—which means homo sacer is outside of the laws that govern proper citizens, whose life is protected by law. But, interestingly, this exceptional position, this life that is outside of society, is itself that which, through marking it as exceptional and other, is how sovereign nations define themselves.[27] Well, this is according to political theorist Giorgio Agamben.

Homo sacer is for Agamben in its exemplarity for political relations the opposite of the sovereign head of the state (i.e., the king).[28] It is the definition of the sovereignty of the king, as the basis for the sovereignty of modern nation-states and their citizens, that is of concern to Agamben. Sovereignty, the supreme authority, is to decide on the state of exception, which is exemplified by the figure of homo sacer. This definition of sovereignty is indebted to a view of the world where nations are all in relation and are either enemies or friends.[29] Homo sacer, as exception, is related to the notion of the enemy because they both can be killed.[30] But, while the enemy can be killed, only homo sacer is removed from all relations as an outcast to whom the law no longer applies. Enemies differ from homo sacer; otherwise they would be the source of sovereignty. The enemy already has sovereignty but one that needs to be destroyed. Thus, while the enemy and homo sacer are related through being killable, Agamben's introduction of homo sacer points to an exceptional positioning in a world made up of sovereign nations, a position that is radically outside of this world: "Political sovereignty entails the legitimacy of killing—not as an end in itself but in order that a greater life, that of the polity, can be protected and maintained."[31] Hence, the political is defined as a particular relation to the right to kill. While killing an enemy is the protection of sovereignty from usurpation by another, killing homo sacer is the image of protection as the claim over the lives of proper citizens.

How does this relate to the situation of Black lives? How does this relate to the question of identification, border patrolling, and the outsider status of blackness from relations among those who have a standing in society based on their gazing that opposes? While the figure of homo sacer provides us with many similarities to the situation of Black lives, it also differs in one important respect: its temporal narrative. Homo sacer used to be a proper citizen. They lost their claim to being treated like a proper citizen. But after Fanon, this is only partially true for Black lives. While Fanon does also bespeak such a loss through his narrative form, where Martinique was an uncontested belonging before, *such narrative breaks in the very moment, in the very temporality announced by (revealing of) the white gaze*—because the white gaze is always against Blacks. In the case of Black Swiss, there is no before, where they owned citizenship as Blacks. Their citizenship is always only partial. The narrative of being able to join proper citizenship, as the joining of this gaze, as an attempt to leave blackness behind, nonetheless always fails because Black lives always fail in properly joining this anti-life, antiblack, gaze, exactly because Blackness has no resistance. Black lives cannot escape blackness, like the ruse that non-Black lives possibly can, through locating blackness at the borders, in those who are Black. Black lives, even when erasing blackness, are always carried away by blackness, as they have no power. But this is an inhibition of lives; this erasure of blackness means structural violence against Black lives and lives close to black; it means being continually subjected to identity searches despite being Swiss.

Let me wager that blackness is in the position of homo sacer before ever being able to become homo sacer, as they can never fully join proper citizenship (and here, this "before" marks not a temporal movement but a positioning). It seems that homo sacer also distances the exception (as exception) from that which it is exception of.[32] By making homo sacer a state that is marked by loss, citizenship is proclaimed as already there before anything else. In this sense, this loss protects this kind of life from the possibility of having been something else before. Since homo sacer is the origin of it, then if that which founds it is after it, then it must, if it wants to be an authority over homo sacer, erase the possibility of having been founded. If sovereignty is absolute authority, then it could not be made by anything or anyone else, because that would make such have authority over it. But blackness thus seems to pose a threat to this claim to absolute authority. If blackness is indeed not locatable as mere loss, as only after this authorized life as a citizen, then it undoes this absolute authority.

Well, actually, I must clarify: homo sacer "presumably" didn't even lose his citizenship.[33] Thus, this proper life, this life authorized by (and as) sovereign authority, is doubly protected. Not only is the loss of a position in this life always after having been in it, but *it is also always only partial*. The outside of the state of exception, of homo sacer, is always partial. Here it is close, in an inverse relation, to my elaboration of antiblack color-blindness, as the erasure of blackness, as being able to be a citizen but only partially. And what is revealed is that this distancing, this question of either you're a national or black, is itself analogous to this structuring principle of claiming absolute authority over life, through the right to kill.

But the similarities don't end here. Homo sacer is also a border zone—the space between proper life, civilized life, and basic, or *bare*, life (also thought as animal-like life). While proper life is civil society, made up of citizens, bare life, also called "mere life," is that which is absolutely outside of such proper life. Homo sacer is the negative in-between of these two because he is life but life that cannot be sacrificed, like bare life, and life that can be killed, unlike proper life. It is here in the separation between them that citizenship is joined, but it is joined not through this separation but through the subsequent erasure of this in-between space through the figure of homo sacer, as a border zone, of which the underground is mere life. This is why Agamben says that it is in this border zone, in this interstitial, marked by homo sacer, that the separation is not only articulated but also where these separated lives can become "indeterminate" from each other.[34] But this fact, that the lack of separation is also the articulation of its absolute separation, can only come to the fore when thinking through blackness, thinking through the exception's nonexceptionality. The reason for such is that citizenship, if it must be joined and is not pregiven, requires itself both an opening of a gap between mere life and proper life, and then the subsequent erasure of mere life from the space of proper life or citizenship.[35] While homo sacer is a border zone, just like all questions of race, proper life is color-blind and antiblack; it is an erasure of the borders, or changing, of life. Homo sacer is the border zone, where absolute control over the outside is exerted through the fashioning of this in-between space, between proper life and bare life, that also allows for both an absolute erasure, as the dispelling of the outside, and an erasure of their separation, also as a dispelling of an outside—the claim to absolute authority, demonstrated in being in control over borders.

It is the introduction of bare life into politics—the politicization of bare life as such—that marks modernity for Agamben, of which the Nazi regime is for him *the* example.[36] Thus, Agamben theorizes from a ground where *proper life* can be in the political without *bare life* present (although it forms an absence in every sovereign), and it is this conception that allows him to see an entrance of *bare life* in the atrocities of World War II. But if, indeed, citizenship as sovereignty is made in the move from bare to proper life as the control over the in-between, then citizenship is exactly marked by this erasure of bare life from within itself, as its own distancing from this border zone. As Samuel Weber, for example, points out in relation to Agamben's insistence on an emergence of a politicization of bare life—which is also to say, an absence of bare life in the political before his matter of concern in the analysis; namely, the camps of World War II—this does not explain why *certain* people have been grouped together to become homo sacer within the camps (i.e., racialization as the grouping together of peoples based on attributes).[37] What if the very illusion of absolutely separating proper life from bare life is what allows for the very conflation of the two, where bare life as one kind of biological life becomes proper?

Others have voiced the same criticism as Weber;[38] for example, Willem Schinkel, who brings to the fore how there is a stepladder according to lines of proximity to citizenship.[39] Furthermore, this hierarchical ordering of society is a question of what proper life makes as the norm—more exactly, how traits are identified for groups that are supposedly outside the norm, which erases the very norm inside, but also in turn conflates a singular attribute with proper life by leaving it unspoken. This does not mean that these traits that are marked as other are more studied; in fact, they are marked by being stereotyped as representations of the unknown, an unknown that must continually be held off. Foreigners are positioned differently in relation to the inside, depending on their distance to the border zone (similarly to the example given of Switzerland in chapter 2). Concludingly, I read here a spatial distribution of citizenship, not according to geographical location alone but also as a function of proximity to that which the sovereign state *is*—is made by and is protecting; namely, citizens and citizenship (and of course land).

The question of biological control is also crucial to philosopher Michel Foucault's initial development of the notion of biopolitics, in whose footsteps Agamben follows with his elaboration of homo sacer—biopolitics as the study of the way in which life itself becomes subject of politics. As Schinkel points out, Foucault initially developed biopolitics in relation to population control via techniques such as measuring birth and mortality rates.[40] At the same time, and with such techniques, biopolitics is concerned mainly with the milieu of the human species.[41] Both Agamben and Foucault stress a spatial character of biopolitics but into different directionalities: Agamben stresses clearly that which lies outside of politics, of civil society, as state of exception, and Foucault the outside of *normative* society, bespeaking the "shaping of a population," through the exclusion of certain lives and not others on the basis of race.[42]

The tools that exercise control over what is normative are "the statistical delineating of populations, the identification of risks, the reduction to individual 'cases' and the identification of 'danger zones.'"[43] These remind us of the situation of Blackness in Switzerland: how population measurement is purposively used to erase Black lives from within, how risks are identified in the form of Black bodies as well as kinship relations outside of the proper, and the reduction to individual cases, where antiblack racism is removed from society and its structures itself. By way of this analysis that places population control within biopolitics and, as an extension, the normalizing of what a society is, biopolitics is tied into questions of ethnicity and race (and the protection of society along lines of kin).

There is another dig we might join in on, one that also uncovers a real historical figure that may aid us in thinking through the situation of Black lives today. Black studies scholar Alexander Weheliye similarly points out how Agamben's theorization of biopolitics is also, despite it not being figured as such by Agamben himself, racializing, evinced in another figure Agamben invokes for his theoretical elaborations—namely, the *Muselmann*. *Muselmänner* were those detainees in Nazi death camps who suffered from such "chronic malnutrition and psychological exhaustion" that they resembled "corpses."[44] As Weheliye describes:

> Due to extreme emaciation, often accompanied by the disappearance of muscle tissue and brittle bones, the Muselmänner could no longer control basic human functions such as the discharge of feces and urine and the mechanics of walking, which they did by lifting their legs with their arms, or they performed "mechanical movements without purpose," leading the other inmates and later commentators to view becoming-Muselmann as a state of extreme passivity. Observers portray Muselmänner as apathetic, withdrawn, animal like, not-quite-human, unintelligible—in short, as ghostly revelations of the potential future fate that awaited the still functional inmates in an already utterly dehumanized space where everyone was exposed to chronic hunger

and death. Being forced to occupy a phenomenological zone that could in no way be reconciled with possessive individualism, the Muselmänner exemplified another way of being human and were, in fact, likened by several observers to starving dogs.[45]

Two things are noteworthy to point out. One, the name *Muselmann* is "an antiquated and now derogatory German language designator for Muslim men that was also applied to women";[46] and two, how *Muselmänner* were placed outside of the category of human life. For Agamben, the *Muselmann* marks a "transcending" of race because of the reduction to the barest of life— an "absolute biopolitical substance."[47] But, as Weheliye notes to the contrary, the *Muselmann* represents "an intense and excessive instantiation thereof."[48] Firstly, if racism still exists, then how can it do so without producing the categorical distinctions it requires (i.e., race); and, secondly, if race is not biological or cultural but rather politically and legally produced and reperformed, then "the biopolitical function of racism is race."[49] Weheliye then expands his elaboration of the way in which biopolitics is racism and not outside of it by engaging Foucault's articulation of biopolitics, which does start with race but still places historical racism outside of Europe, as a matter of the colonies only.[50]

Weheliye elaborates extensively on the relation between the Negro and the *Muselmann* as it pertains the articulation of race within biopolitical techniques.[51] As Weheliye concludes in his elaboration of biopolitics and its entanglement with racialization:

> What connects the classifications *Negro* and *Muslim*, then, is not biology per se but their conscription to a set of political relations that necessitates inventing new caesuras in order for Man to remain interchangeable with the human and that these relations are sociogenically imprinted to generate hieroglyphics of the flesh. In the United States the Negro came into being when the slave no longer accomplished the required labor of distinguishing black from white subjects so as to ensure the continued superiority of Man with its attendant class privilege, at the same time as *Muslim* became necessary as a racialized category in Europe when it threatened to dislodge the until then unchallenged advantages of whiteness, Europeanness, and Protestant secularism of the autochthonous population.[52]

Weheliye poignantly demonstrates that race is not reducible only to phenotypical features but is articulated—that is to say, created at and as the borders of proper life. On the other hand, our elaboration in chapter 2 problematizes a certain tendency toward grouping those who fall into the racialized category together, such as the foreigner in Europe, because such can once more be used to evade the question of race and blackness. It is indeed true that the place of blackness within the hierarchy of racism can be ascribed to other groups, yet those who are placed into the category of blackness through biological features or an absence of proper blood-kinship are not able to ever fully shake this blackness.[53] Additionally, we have to point to the issue that even within the scene of the Nazi death camp there is Blackness as absence present. (As many accounts in *Showing Our Colors* document, Black Germans were *also* targeted by Nazis.)[54]

It seems that citizenship itself is formed by antiblackness as the process of placing racialization and race outside of its *space*, into its border zone, and claiming control over such. The question arises as to whether that means that Blacks are, as a result, in the position of noncitizenship. This is exactly the question Christopher Busey and Tianna Dowie-Chin engage. They point out that citizenship is tied to both racialization and questions of belonging on a global scale. That is to say, that nations form the global and citizens form nations—the global is articulated through nations as citizenships. If the world is imbricated in antiblackness, whether that be thought through the globality of the color line or the importance of coloniality for the world as we know it, and such world is formed of/as nations, then the question arises as to what blackness's relation is to nations. For Busey and Dowie-Chin, the world's antiblackness points to blackness as being positioned as anti-citizen rather than noncitizen. As they poignantly observe, the notion of noncitizenship implies that there is some elsewhere that the noncitizen belongs to, whether that be in terms of geography or biology, and/or that citizenship can be achieved, and I would add lost, at some point in time.[55] As they explain:

> Anti-citizenship is thus more appropriate because it implies direct resistance to or being (as lived) against the state and never part of it. Simply put, the Black subject is the ontological other of the Euro-Western nation, "the scapegoat for White society . . . [or] oppositional brute force." Consider that in curriculum, Black subjectivities enter the cognitive sphere as either primitive or in the non-human status of enslaved persons. Black subjectivities are also positioned as inherently transnational, as human cargo primed for global dispersal within a capitalistic system of exploitation that is only implicitly named. Any

effort to have Black humanity recognized is always presented as counter to the state, hence the prefix anti-.[56]

I'd like to build on and depart from their definition, which is also to extend anti- with ante- and to shift the meaning of such prefixes. The clarification requires a rethinking of the resistance's placement away from a push against to a push that is revealed within that which is prefixed: prefixes become revelators of hidden, sedimented meanings within a term. *It is not simply that Black lives are anti-citizenship as another kind of outside (albeit an absolute one) that could be conceived of after the fact of citizenship (that is merely produced) but rather, as our elaboration prior aids us to come to grasp with, blackness is a before (that is, it is before the world as antiblack, before the citizen but placed, causally, outside, and thus also after, the nation).* Blackness is the underground of the border zone, where life is articulated, but also where through the attempt of erasing this very border, this moment of separation as the claim to authority over it is born as the attempt to make citizenship an always already given proper life, in contradistinction to a bare life. Citizenship's antiblackness is there because of its own erasure of itself, of its sound, through erasing how it changes beyond its control. It is through the distancing from (and owning of) blackness that it tries to gain its absolute authority, as the fabrication of borders, as something that can be owned and managed by it. In some ways, a reading of Achille Mbembé's notion of necropolitics, as the flip-side of biopolitics, as the politicization of death itself, can aid us in elaborating the way in which this erased world, of borders as ends, assumes its very quality, or name, in this claim to authority. We might also say that in sovereignty's fear of not having absolute control over life, over what constitutes proper life, which is expressed in the claim to the right to kill, comes to the fore also, how it must create death and death-worlds.

Death

Busey and Dowie-Chin point to the question of the global as another border-making, where the world is made with and through the forming of sovereign nations. Achille Mbembé departs from Hannah Arendt's work *The Origins of Totalitarianism*, where she demonstrates how the making of the world and European sovereign nations required an uncontested space, a space where laws did not apply: Africa.[57] Mbembé's theorizing furthermore departs more directly from Foucault, specifically from Foucault's initial development of biopolitics as directly informed by questions of racism. As Mbembé states:

"In the economy of biopower, the function of racism is to regulate the distribution of death and to make possible the murderous functions of the state."[58] Mbembé's theoretical elaboration flows from biopower and exceeds it by moving into another *space* of focus under the heading of necropolitics, specifically by way of an attention to the question of the distribution of death. Mbembé summarizes: "The notion of biopower is insufficient to account for contemporary forms of subjugation of life to the power of death. Moreover I have put forward the notion of necropolitics and necropower to account for the various ways in which, in our contemporary world, weapons are deployed in the interest of maximum destruction of persons and the creation of *deathworlds*, new and unique forms of social existence in which vast populations are subjected to conditions of life conferring upon them the status of living dead."[59]

But like biopolitics, Mbembé needs a definition of what constitutes death and life: "Death is the putrefaction of life, the stench that is at once the source and the repulsive condition of life," and the sovereign transgresses the "natural" prohibition of death.[60] As Mbembé elaborates, by treating sovereignty as the violation of the prohibition, the question of "the limits of the political" is reopened.[61] Thus, "politics, in this case, is not the forward dialectical movement of reason. Politics can only be traced as a spiral transgression, as that difference that disorients the very idea of the limit."[62] In other words, politics requires the fabrication of a limit (a border zone) that continually must be transgressed by it so as to claim authority over this limit, which is also what makes the limit into the limit (i.e., places it under this name). Hence, similarly to biopolitics, there is a sovereign space that is bordered by the state of exception, although this time this is not a question of a right to kill as a marking of the limit, but death itself as limit whose transgression, the moving over this border, demonstrates and inaugurates sovereign power.

In this analytic, the example is given by another scene; namely, the plantation.[63] Within this scene we find the slave who is marked by a triple loss: natal alienation (loss of mother, blood, and kinship), absolute domination, and social death.[64] In this articulation of life and death, the slave's life is stolen by the master, and the slave is marked by loss. Here, the question arises: What did the slave lose if the very losing is what marks the incipit of sovereignty, which in turn marks the notion of life and death?

Necropolitics as an analytical tool allows us to uncover the political limit—that is, also the limit of sovereignty and citizenship, in a direction of a previousness and an outside, which sovereignty makes (and transgresses) to claim itself. This direction cannot be discussed in biopolitics because biopolitics forgets the ground on which it stands, the limit, which is placed somewhere else than Europe, in the colonies, in the plantation, in the space of blackness, in the border zone. In this way, it is closer to our aim to elaborate on the position of Blackness within the scene of citizenship. But it seems that necropolitics remains within the bind of seeing the limit as outside of citizenship. Or, we might say it in this way: it seems that the inside of citizenship remains absolutely separate to this outside that is the space of death. Proper life is separated from death, and thus necropolitics bespeaks the situation from the perspective of an antiblack world, where death can be cut off, where the limit can be held off, so as to be able to transgress it, exercise control over this border. In this sense, the space that is wilderness and death, as source of life, used for sovereignty is located somewhere else—that is, it is spatially and temporally removed from civil society, and we theorize from within.

Similar to Weheliye's point in relation to biopolitics, where the *Muselmann* demonstrates how race is politically formed and then ascribed to biological and cultural features, so is hence Mbembé's work pointing to the fact that race is politically constructed and then ascribed to features and creates a space of death as separate from proper life. Here, on the other hand, the racializing process of the political is center stage, but there is no way to account for how such features are chosen, except for historically, which then in turn must continually search for an origin in historical conditions.[65] Hence, we have to pose the question of how the historicized positioning of the Black as the bottom of the latter can be accounted for within the discourse of citizenship and sovereignty, which is not simply a matter of historical elaboration but of an order of a problem that announces the historical itself. In other words, any historical account is tied to questions of sovereignty in the form of authenticity as well as in the emergence of history as a study rooted in the founding of modernity. That is to say, the authenticity needed for the proof of racialization is itself thus tied to the sovereignty that gives itself authenticity. Antiblackness turns out to be a self-reifying system where even racialization, and necropolitics, even death, are produced on the basis of its own authority.

Mbembé is himself laying out a series of historical arguments within the development of necropolitics that mark the Black as having a particular relation to necropolitics or the ground of sovereignty as death.[66] But if, after Mbembé, the fabrication of death marks the origin of sovereign life, or life as citizen, then death is *of* sovereign life. That is to say that in the notion of death as limit for transgression lies an unarticulated claim to authority over all life, which must also make (or we may say own) death so as to be able to transgress it. In other words, the act of transgression is a narrative form that proclaims itself in the very claim/act of transgressing (maybe we can even say, in naming transgression).

Mbembé's elaboration on Paul Gilroy's reading of the act of suicide by slaves as an act of resistance to being put into the position of death is telling here: suicide by slaves was an act of release from bondage.[67] Here, Mbembé also points to a temporality within the structure of freedom that is of a necropolitical configuration where release, or freedom, is found in death itself. It bespeaks the structure of freedom itself: "The future, here, can be authentically anticipated, but not in the present. The present itself is but a moment of vision—a vision of the freedom not yet come."[68] It seems that the right to kill is accompanied by the invention, as the definition of not only life but also death. It is the claim over the definition of death that allows for the claim over (the definition of) life, as also the claim to having (a) life, which is denied to death and those who have no life, who are *close* to death.

At once Blackness is living death and bare life—life and death as before and after life and death, where life is previous and animal-like (bare and stripped of its value) and death is after transgression (it is always already transgressed). The right to kill Blacks, and blackness, is the right to kill death itself, is to kill that which is ascribed death. Countable, proper life is formed through stealing it from Blackness in the first place. In other words, by stealing life from bare life and then erasing such a steal, erasing such beforeness of bare life, proper life is claiming its authority over life. This is why the Black is in the position of gratuitous violence: "Whiteness, then, and by extension civil society's junior partners, cannot be solely 'represented' as some monumentalized coherence of phallic signifiers but must, in the first ontological instance, be understood as a formation of 'contemporaries' who do not magnetize bullets. This is the essence of their construction through an asignifying absence; their signifying presence is manifest in the fact that they are, if only by default, deputized against those who do magnetize bullets: in short, White people are not simply 'protected' by the police, they *are* the police."[69]

This elaboration by Frank B. Wilderson III reveals not only that civil society is a deputization against blackness, "those that attract bullets," as a matter of protection for society and concurrent joining of citizenship, but also that those who attract bullets are outside of that which is protected *as* civil society. That is to say that if civil society is marked as proper life, then the biopolitical and necropolitical enforcements over bare life are simply the way that civil society unmarks itself, trying to claim absolute authority. Thus, proper life, the life of a proper citizen, is not really already there; rather, it requires the *putting to death* of bare life—the delimitation of death and its transgression both at once, as the claiming of the right over death through the right to kill. *Such right to kill, or the power over bare life through making death, through the right to kill, is the very fabrication of the separation between (and the making of) proper and bare life itself.*

This is undermined by the state of exception's apparent exceptionality. That is to say that in the claim to exceptionality, as something that may happen to an unlucky few, as source of sovereign authority, lies a denial of the exception's abundance: in other words, the exception has to be exceptional. Here we can pair exception and transgression, because like transgression, exception tries to erase its relation to that which is the exception. And, like exception, transgression tries to erase that which is transgressed through claiming that it can be separatable from it. In the notion of the transgression of death as sovereignty is an understanding of life as proper life. If proper life is not before the exception and it is enforced through necropolitical manufacturing of death, then politics is made in death—namely, through the identification, as the definition, owning, and administering of death.

This is evident if we think of the very tools that are used to claim proper life, such as, for example, the control over populations through statistics. It is only through death that life can be counted in an absolute manner—life makes its value, as proper life, in its orientation against death. It seems that the control over life happens through the attempt to own and manage death through administering/making it and thus managing who, or what, gets to be a life. Strangely enough, it is the very excessiveness of life in bare life that frightens Agamben, that to Agamben life in general/bare life can become more important than individual lives. While for him it seems that such takeover is evadable in pure proper life, we must problematize such. For there is no such thing as a proper life without bare life. And even worse, it seems that proper life itself, as the distancing from bare life through the erasure of its separation from it, is the very takeover of life in general. This is how Samuel Weber explains it:

> If the notion of "bare life" could become, as Agamben argues, a lethal machine, it was in order to purge life as lived by singular living beings—which entails the constitutive relation of life to death—by subordinating life in the singular to life in general. Life in general can be seen to generate and perpetuate itself prior to and independently of death. Life in the singular cannot. But life in general is only life when it assumes a particular, nameable, identifiable form. The name that this form assumed in the modern period was most often that of the "people"; in

particular, a people unified by ties of blood and soil—the biopolitical version of the criteria of modern citizenship, birth, and territory.

But a "people" is inevitably particular and limited. So the question emerges as to which "people" is to be regarded as the decisive embodiment of life in general? If the essence of such life in general is its capacity to kill death—that is, to expunge death from life by eliminating all those who embody its irreducibility—then the status of a people to defend and protect life in general will have to be demonstrated by its ability to kill and eliminate its enemies—who by definition can be considered to be enemies of life.[70]

To attempt to kill death is thus what Wilderson's citizens are doing when they kill blackness. But this is not only to attempt to kill other people but to kill the possibility of death itself. If making death so as to kill death is necessary to make civil life, then what is death other than life as proper life? Weber alludes to how mere life escapes the confines of death, as it cannot resist, nor experience, death.[71] This inability to resist death is also the condition of those who come to the fore in necropolitical analysis: blackness cannot resist death (just like Blacks cannot resist the white gaze). It is an excess of life that can be found in the space of blackness.

There's a flip side to all of this. If proper life as proper citizenship makes itself in the claim over borders, then blackness is a kind of underground of the border zone where the border zone is not a border. What if the notion of the border zone is the erasure of it, or that which it carries in it, through a claim to having authority over who gets to cross the border? Or, to put it in different terms, *the very articulation of race is already the erasure of it and its underground: blackness.* Because if race is about purity and if impurity becomes that which it manages by policing and making it into a border zone, then behind the term *race*, or *border zone*, are two things: one, it is a cover-up to erase something, also turned negative, that is a kind of life that is not only not normative but from which the inside of citizenship both protects itself but also claims its omnipotence in *opposition* to, i.e., blackness. And two, behind race, or border zones, as negative identification lies life's liveness, how it is born and dies but never ends, not because it is some kind of endless cyclical thing but, rather, because life is being articulated, life continually is called in, called in in its own notness as a relation where borders are crossed.

But what does that mean regarding citizenship? Is citizenship tied to proper life and sovereignty in the form of always opposing and making border zones as separatable from it? Or maybe the question requires a change itself: What if it is about recognizing how citizenship is full of border zones and that the policing of such has become the technique of erasing that which is their underground (i.e., an excess of relations across)? It seems that all the antiblack methods of policing erase border zones, or make border zones, as violently contested nonspaces, somewhere far away from the inside of citizenship and the nation. The identification of blackness as the erasure of it from the inside through making border zones to it for the creation of a proper life seems to be claiming ownership over borders—that moment, or nonspace, or singularity where proper life articulates itself, through the distancing from them. By pointing out a bare life as its opposite, as its exemplary exception—that is, both where it comes from but also in the same term and under the same name, erased—proper life makes itself into the proper. Thus, rather than thinking bare life as some real thing before proper life, or proper life as real or illusionary itself, we must come to grips with how proper life erases its own bareness through the identification and delimitation of bare life, as something unchanging and outside of it (death-worlds or endless afterlives). Blackness bespeaks an excess of the border zone, as that which is not its name, not that which identifies it in a world made of proper lives. Blackness refuses its identification that is really a disidentification by both identifying itself before and after the proper and undoing its name by naming and unnaming itself in the same turn. Thus, thinking about borders can become thinking about how we name, or make, us. What does it mean to be Swiss? What is life? This is articulated here, in this underground of the border. When speaking up against antiblackness and policing, Mohamed Wa Baile and many others are thus radically pointing out how Swissness makes itself in opposition to its other and how it identifies its border—how borders are identified as a dangerous closeness to Black and blackness, and even equated with them/us. The proliferation of border zones must be flipped on its head, into the proliferation of lives changing—changing what Swissness, what "citizenship," means, what us means.

5

Interstitial Listenings III
Black Music behind the Wormhole

Sonic S.cape
Maïté Chénière's work *Sonic S.cape* speaks of an outer space flight like our interstitial listening that flies beyond holds as secret space travel technology. Chénière, who performs under the stage name "Mighty," was born in Paris in 1992 and has lived in Geneva, Switzerland, since finishing her studies there.[1] She organizes club events specifically centered around blackness and queerness, creating safe spaces for black and queer life, turning the dance floor into "a space for all types of bodies and identities, a platform for expression and togetherness."[2] Created for a dance floor, *Sonic S.cape* is composed of sound as well as video and is supposed to be projected and played back while the audience is dancing. We can thus say that the work is meant to be listened to in dance, in movements, as a way to listen to the sonic and visual text. The work begins with a countdown that lines up the video and the sound recording. As a result, it is clear that the sound, even though it aligns with the video, will inevitably have micro variations of relations to the video component. This togetherness in something of an *unplaceable* manner is played out in

this work on multiple layers, from such small micro timings to larger formal and observable layers—such as, for example, how the piece is experienced on the dance floor with surround sound and video projected onto one wall.³ Let me take you onto the dance floor, a spaceship that Mighty invites us into as a space of Black queer life.

The work begins with the image of a whale-like spaceship in some blueish ocean with a solid circle in the center (see figure 5.1). Looped ambient sounds slowly give way to bass and drums in strange temporal relations—the seemingly uncoordinated becomes poetic relations. As the sounds become more temporally complex in relationship, the image changes to a black background and six video frames within this larger frame, the circle still present. One video frame is inside the circle and the others around it. Film footage moves from the inside to the outside and vice versa, frames changing relation as the sound changes from different samples of Black texts as music, all in different time zones, flowing into each other.

Over the course of the work, the frame within the hold of the circle changes its size and placement, enlarging, moving in front of the circle, then behind it, until slowly the whole screen becomes filled with it, opening another set of dimensions, to the back and front. But the hold is not left behind, and

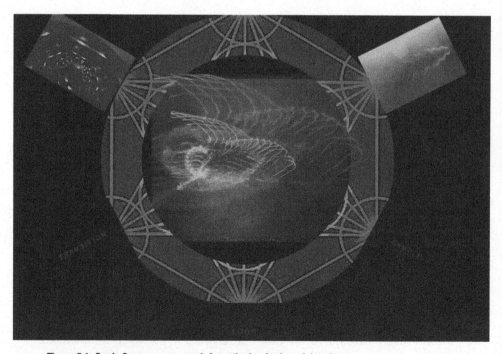

Figure 5.1. *Sonic S.cape* screen-grab from the beginning of the piece.

towards the end, while there is still a frame within the hold, there are lines of light dancing over the screen, like spectral analysis visualizations of sound but somehow moving in counterpoint to the sounds heard—they visually sound like some kind of inaudible components of the music (see figure 5.2). Sounds become a cacophony of musical times, many different tempi move at once, from different songs brought together in something like a chaos: I hear a chaos world of poetic relations.[4] This chaos world, this world in poetic relations where nothing can be apprehended as a whole, where the listener has to dance with different times, shifting place and body in space so as to move into different frames, speaks of a disjointed listening—like an archipelago with different islands and an ocean as where they emerge like waves frozen solid for a period of time. Listening becomes improvisation of new counterpoints to a rich, harmonious chaos of polyphonic music. But it cannot be counterpoint as conceived in point against point, but rather it points out how music always already points in relation—lines are nothing other than the music playing out, in, and with.

At the end of the work, we are left with the circle, the hold (like the belly of a ship), and darkness. The music fades out and we see a pentagram: some mystical alchemical shape that connects the dots previously already found on the circle forming a shape, like stars connected across distant space that, seemingly, was there all along. Just like how all these musical pieces were put together from different spaces and times, having different spaces and times,

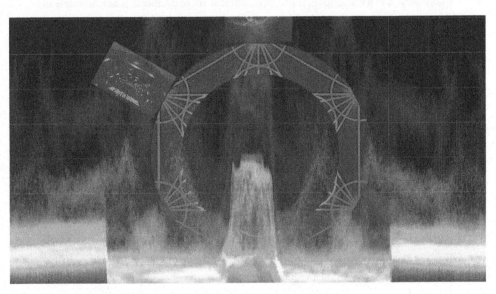

Figure 5.2. *Sonic S.cape* screen-grab from near the end of the piece.

into music, those video frames outside and in the hold are transmutated into a poetics of relation. Thus, the hold becomes a spaceship, a transformational body, a cypher composed for improvisations meeting each other. In the final exaltation, where flesh moves according to its own volition as dancing (no)bodies, holding becomes poetic relations between dots that only exist as knots in lines of relation—not in relation but in poetic relation, in an opaque togetherness because there is no one to be together and no one to be by itself.

The hold is transformed into a poetics of relations, of frames as unframed.[5] There is no frame within the hold and outside the hold; there are only frames as knots of lines of travel. These hieroglyphics, as lines of travel and flight, then don't become but rather they always already mark that which was before/in-front-of borders and spaces they encircle: a chaos world, or an improvisation of points from lines and lines from points, both at the same time, as well as those lines not heard and those points not seen; their dance is an interstitial listening.

Cases, as holds, become transmutated from exceptions, or borders, to interstitial (non)spaces. And the interstitial becomes and exceeds all there is. It is this underground of the in-between—the excess of light dots or what they are there to remind us to see. That empty space forms space and time, and dots and lines. Jovita dos Santos Pinto in *Spuren: Eine Geschiche Schwarzer Frauen* (Traces: A history of Black women) practices a writing of history that I want to read as a *methodology* of "sociology hesitant."[6] She begins the essay with a critical engagement of historical records about Black women in Switzerland. In an archive ridden of Blackness, critical fabulation in all its registers is needed to speak of Black lives (see chapter 1). But ultimately, her historical work turns into an intervention in historiography and it is here that something is clarified about how Blackness shifts the archive.

Her title uses the word *Spuren*: traces, imprints, something left behind by someone who's not here any longer. In the first half of the text, she uncovers traces and what they hint at as untraceable artifacts of Black lives in the scene. For example, in her discussion of Tilo Frey, the first Black woman to be elected into Swiss political office in 1971, the first Black and the first woman, requires a critical reading of the archive.[7] That Frey did not speak about Blackness is recorded in the archive, but what a critical engagement of such archive has to do is think with lives behind the data, beyond what the archive can hold—in the form of Frey's life as well as in the form of Black lives as that which is unaccountable within the archive and those political discourses she was able to engage in.

In the second half of dos Santos Pinto's text, this need for an uncovering of Black lives spills over and shifts the meaning of what counts as archival material for historiography. Dos Santos Pinto declares that since discourses around Black women in Switzerland are only coming to the fore under discourses and narratives that are concerned with origins and migration, which is also to say as not properly locatable as and in Switzerland, it is necessary to also take into account lives lived in this space (of Switzerland).[8] To this aim, she conducts a series of interviews with Black women living in Switzerland and combines those personal accounts with her own scholarly reflections.[9] In this sense, historical writing centers itself around lives lived in spaces and the meetings among Black lives.

Akin to W. E. B. Du Bois's metaphor of the city directory, this shifts the way history comes to matter and plays out by following lives rather than preformed categories of study/measurement. Instead of providing an overarching narrative or searching for origins or validation, this historical project becomes an intervention into the very ways in which history is and can be written or told.[10] This particular practice of *telling* as the study of how to tell, and as such how to listen and speak of and with blackness, of that which is unknown and unspeakable and erased in the current archive, is one that shifts the position of the archive. The archive then becomes a reflection of lives and a practice for them to meet, and Black lives voice themselves. This is a study of history as one that is *open-ended* in multiple directions.

This question of how stories are told about Black lives brings us straight back to *Showing Our Colors*, to Katharina Oguntoye, May Ayim, Dagmar Schultz, and all those lives that are a part of the book's narration in the form of interviews and more. Pinto sees in the network created by discoursing with Black women—the practice of historical writing in which she engages, that is also to say, the spaces articulated in meetings in black—the capacity for political *relevance*.[11] I hear under the heading of women a capacity for politics for Black lives in Switzerland—like a sleeper cell under cover of the fight for women is a fight for women as well as Black lives.[12] Dos Santos Pinto's work points out how within political organizing, under the headings of women and immigrants, a space for Black lives can be created.[13] As such, it is within another container that blackness can be spoken of, and within that, which has authority by being antiblack. At the same time, this needs to be problematized (more in chapter 10). Furthermore, while this, in conjunction with a grouping together according to immigration background, has been the register of blackness within Switzerland as it concerns politics' voicings of Black lives, I want to think in excess of these already acknowledged categories toward an underground insurgence.

What Voting Is For

Maybe there is even another kind of politics announced in this shifting scene, through this steady insurgence work of Black lives and blackness in categories in excess of their formation and part of an antiblack system.[14] If politics is rooted in antiblackness, like the archive, and the archive can be shifted into what was announced under Du Bois as a hesitancy of study, then can politics too? As a result, can then Switzerland, or the nation-state, or citizenship, be shifted as well? Of course, this question must be outside of a kind of proper time accounted for by proper history; there must always already be blackness hidden, as impossibility, as excess of, but right in front of, the antiblack gaze.[15] This question is not raised under the auspices of saving the political or modernity, but rather what we are asking is what is that space of politics that Blackness bears, and how can we follow that which is in excess as not spoken and overtly there. Thus, it is here that an opening for the question of the political in relation to the archive comes to the fore.

Du Bois might provide an incipit for our own refiguring—not because his thought is necessarily the answer for our times and place of concern in this book, but because it might allow us to open our theorizing and study to other ways of thinking about what it means to be part of a nation. And in Switzerland, as a direct democracy, such is of course a very pressing task for everyone who gets to vote, because it is the conception of what is us and what can be us that influences decisions on voting for policies and what is enforced through voting. There are examples of policies that have been instantiated in recent years that demonstrate this question as of fundamental import and entangled in questions of racialization. From the changes to how Swiss citizens are determined—which now also happens through being related to the father and not just the mother, as a positive example—to a more recent, rather troubling one. The latter being the law that inhibits the wearing of face coverings in public spaces, which even in its wording resorts to the term "local customs" as an exception of the application of this law (the other exceptions being health, weather, or safety).[16] Its introduction was intended for Islamic face coverings such as the burka and the niqāb.[17] Issues of Islamophobia with regards to this law and the discourses surrounding it, as well as the entangled question of women's rights, have been raised by others.[18] Here, I would like to raise questions about the occurrence of the notion of local customs that appears in the phrasing of this law. Added because of the need to allow for the custom of *Fasnacht* (carnival),[19] it seems problematic because it attempts to solidify what the local is, and thus what Swissness is.

While this law is geared toward face coverings, it also bespeaks a general issue of attempting to limit what Swissness, what local customs, can be and become. Just like citizenship, culture becomes controlled along borders of proper belonging and is used to demarcate it. As a kind of container of the attributes of proper citizenship, cultural practices become sites for border control and antiblackness. Numerous occasions of antiblack costuming have occurred in Switzerland's carnival.[20] What does it mean that our very cultural practices become border zones, where culture is held in opposition as a border zone for demarcating who is proper and who is not? What happens when the very people who are able to bring life to Swiss *traditions* become excluded from them? What happens when people like Jolo, who teaches whole villages the music that is supposedly not his, are held off from tradition through brutal border controls (see chapter 3)? What happens when the freedom of association, so valued by the Swiss, becomes inhibited by the very culture it is supposed to protect?

Policies voted on by Swiss citizens directly affect decisions of inclusion and exclusion, as well as who gets to have rights and who doesn't. Thus, what if instead of conceiving democracy as a form of border control, that is in service of it, lives in excess of what is supposed to be here? It is by inhibiting the telling of our stories through policing what stories may be told that our story, and us, which is narrative but also materially performed, becomes death.

Du Bois's affirmation of democracy is the conceiving of democracy as derived from the *unknown* "freedom" that is sounded by blackness *here*—that is, lives living in excess of the proper.[21] The bringing forth of blackness as a sticking with blackness—after all, we are not *doing* or *having* blackness because blackness is before a claim to agency, and blackness is always not by itself, but some of us *are* Black and some of us aren't—is concurrent with supporting life in more than all its pregiven and delimited forms. This is why scholar Nahum Dimitri Chandler points out that Du Bois placed emphasis on ideals of life based in this unknown freedom, as democracy:

> Thus, we can underscore that the most general and singular concern of the work of Du Bois across the entire itinerary of his thought was the construction or reconstruction of what he called "ideals of life," those headings of value and distinction that would orient a collective social life, the terms that would assist in the organization and sustenance of

a collectivity. While such terms would give the social and historical space for individuals to realize themselves, they would, above all, be operative at the level of the group (humans in general, civilizations, "races" or cultures, nations, and states, or a political entity as such). This general concern yielded a certain practice of principle in Du Bois's thought: an affirmation indeed of the possibility and authority of truth and law as a guide in the organization of life but, equally, a resolute and unfungible affirmation of *freedom or chance* in human doing. In fact, the latter might be understood as the root possibility of the former in Du Bois's terms.[22]

Ideals of life as a practice of living together, to support life, is to live with and for each other. This is why these ideals of life are not simply about that which sustains basic biological needs, nor about rights tied to the protection of proper civil life, as rights that are against those who cannot enter citizenship or proper life.[23] To affirm democracy as the support and forming of ideals of life is to hear one's life lived in relation to other lives lived in excess of proper being and proper relations (i.e., whether known or unknown, whether here or there, whether then or now). And emphasis must be placed on the uncertainty, the unknown, that comes before any laws and any authority, before democracy as that which *informs* democracy.

In Switzerland, this thinking with blackness and Black lives includes thinking about racism, includes thinking about refugees and foreigners and how they're put into a border zone where they cannot live freely: they cannot travel, vote, study, or move about without hindrance, profiling, and so forth. Blackness pulls at the seams of the antiblack world from within and without because antiblackness is the belief that one can be by oneself (even though you must end). The antiblack gaze is a color-blind gaze because it is not allowed to see anything related to black, and with such it must also make blackness as abject—place it and hold it at its borders.[24] Thus, the antiblack gaze, the white gaze, the gaze of an order of politics that holds off the very changing and living of lives, is undone in blackness as an impossible sight where seeing and the self break open. Listening turns borders into interstitial black study projects, as celebration of our world in poetic relations. Let us dance to the chaos world that is our Afrofuturist outer space flight as the (re)writing of us and our stories.

These interstitial listenings are transmutative. By transmutation, I mean how blackness, which was placed as the abject, is revealed as that which is lives lived in excess of a delimitation that opposes and attempts to own death.

Thus, to be with blackness is to be together in music, together in sound, even when not together, both at the same time, because togetherness cannot, here, imply beings reduced to one among others that meet. Rather, Blackness is that which speaks of a together-apartness that is before and out of which all singularities come from—it is radically before any togetherness in contradistinction to apartness. It is like Mighty's music. This book is not by itself, even when it appears as such: it is always with those who are named within it and those who came before it, but also those who touched my life, the lives of the readers, as well as those yet to come, yet to be unearthed in it, who will shift the timbre of my voice. Blackness bespeaks an incompleteness theorem, that asks us to keep digging, as a continual reworking of our stories and us.[25] It is not that the future is radically open, marking something that is closed in the past or present—rather, everything is always open, even in its closedness. How otherwise could infinities calculate into singularities from nothing, like in Jérémie Jolo and in Chénière's musics?

6

Mothership Connections

Opening Space

As we've come to understand through our elaborations, border zones are not only geographic but also biological. Border policing in the biological is enforced on the skin and through blood, through kinship. When listening to Charles Uzor's score for *Bodycam Exhibit 3* while thinking about the task of this book, George Floyd's screams during the moment of his murder, "I can't breathe ... Mama ... Mama ...," while police kneel on him while on the ground, continued to ring in my ears. It is a moment where both motherness as well as its denial and the policing of it in an antiblack world are evinced. "All mothers were summoned when George Floyd called out for his mama," said Valerie Castile, who also lost her son, Philando Castile, to police brutality, in an interview after she watched the video of Floyd's murder.[1] As Jennifer C. Nash explains in *Birthing Black Mothers*: "Her tragic statement exposes the repetitive and seemingly endless nature of antiBlack violence, and it also reveals how Black mothers have been positioned and have had to position themselves to garner political visibility, as always proximate to tragedy grief."² But, as I uncover in this chapter, it is in the thinking and standing with Black mothers that a radical protest and refiguring of the world comes to the fore.

In *Showing Our Colors*, this radical work spearheaded by Afro-German women, another instance of how the policing of motherness is a question of proper belonging and antiblackness is revealed. While this example is of Germany in the 1920s in particular, it points nonetheless to a more general and still current question. They explain:

> The Society for Racial Hygiene (founded in 1905) began conducting sterilizations for "eugenic reasons" in 1919 for the protection and elimination of "racial diseases." In 1927, the commissioner of the Palatinate informed the Imperial Bureau of Health that considerable cause for concern would arise as these Black children matured. He inquired whether it was not possible to render those children infertile through a painless operation when they reached puberty. At that time there was no legal basis for such medical intervention, and restrictions governing sterilization stated that illegitimate children, who generally had the citizenship of their mothers, required a parent's consent for any such intervention, and only in very rare cases, under duress, was it given. However, it is difficult to prove at this late date how many children were secretly sterilized or allowed to disappear, or who grew up in welfare institutions rather than under the protection of family or other caretakers. In the mid-1920s, the Imperial Ministry had already considered handing over "the half-breeds" to mission societies, with enough financial support to send them abroad. Quite apart from that, it was easily possible to sterilize children under the guise of preventing a "racial disease," as long as no one had them in protective custody.³

Underneath the border zone of who can belong—the brutal delimitation of kinship and who may live in a space—is motherly agency. Motherness, as a question of how we make relations, provides a site that can break the delimitations of life and of belonging. Blackness opens space with the mothership.

Black Kinship

If slavery is a moment in time that stands as a metaphor (a very real one) for the inauguration of the modern nation-state itself, then we must think with the experience of Negro slaves. By this I mean not only thinking through brutalities enacted on Black bodies but thinking about a Black radical *livity* (it is more than a practice) in spite of it. What emerges behind the smoke screen of border zones of proper modern life, behind that which cannot see and shall not be known, are alternate practices and experiences of belonging, relations, and kinship—not as an alternative but as a radical (and we might say, continual) refiguring of the terms of life and our world.

I would like to engage, to this aim, first, the example of *Betty's Case*. Scholar Sora Han elaborates:

> In 1857, Chief Justice Lemuel Shaw of the Supreme Court of Massachusetts declared Betty to be free. Betty, a slave woman, had been brought from Tennessee into Massachusetts by her owners, the Sweets, and by virtue of their travel and stay in Massachusetts, the Sweets' relation with Betty had been legally converted from one of enslavement to one of labor. It was within this latter context that Shaw determined Betty to be a contractual agent with free will. This case, driven in its ruling and circumstances by a question about the legal personality of a slave, would come to be called *Betty's Case*.[4]

What marks this case as exemplary, according to Han, is that Betty decided to go back with her owners despite being declared free and, hence, nonperforming the court's ruling—meaning she did not *perform*, or follow, what the court asked of her, which in this case would have been to be free and not return to Tennessee with the Sweets.[5]

Betty's Case marks a resistance to freedom and slavery through a refusal to accept a freedom shaped *by* slavery as the foundation of the modern world—a freedom that means entering a proper life, a life that is rigged against her in the first place. This case isn't about a kind of "voluntaristic theory of enslavement," but rather about how Betty's choice exceeds the very structures of law and citizenship rooted in the relation among proper individuals, or subjects of a state, that is, citizens.[6] In my view, there is, here, another cover-up of a kind of interstitiality through a negative term. In other words, the very term *nonperformance* marks something uncontrollable and thus, for a world where nothing can exceed the authority of the state, a negative. Just like turning *black* into a negative, accompanied by a border for absolute control, where this negative becomes properly managed, nonperformance manages, as a term, in this case, the absence of law's inability, as the state's inability, to exercise control over Betty's choice. Thus, while I follow and agree with Han's argumentation around *Betty's Case*, I would like to bring to the fore how the very term is an act of masking, and thus that this masking also reveals something about law and the enforcement of citizenship itself. It reveals a hole in the middle of a system, or a world, that claims to be applicable to all things, and all-knowing, as well as unchanging, except for a kind of change that it authorizes (such as a notion of progress). It is, in this sense, not about a cover-up of some kind of truth but rather of a lack, which turns into lack only as the absence (or excess) of proper relations.

The story Betty should be telling is one where being a citizen is leaving blackness and slavery behind for freedom. But she refuses both the name's meaning and the rescue from it. Black becomes something that exceeds what this name means for the world of men—men as of equal status so as to enter relations of exchange, to enter into contracts, to be a part of a social contract.[7] By stealing away, covered by an unknown space called nonperformance, Betty is continuing to practice the kind of life that she, demonstrated in the same turn, always had been living despite, and beyond, the imaginings of these nations that make themselves on the basis of owning slaves. This does not mean that slavery is not brutal, nor that Betty had it *good* with the Sweets. Rather, her nonperformance as an exception must be thought in excess of the exceptionality and temporality of nonperformance as a cover-up of a lack in a supposedly all-encompassing and omnipotent system of justice.

Freedom and slavery, and the emancipatory move out from slavery into freedom, are marked by slavery. In Betty's case, she nonperforms her lawful contractual obligation to be free by returning with the Sweets, effectively breaking the contract. In this (non)act, in this nonperformance, Han locates a refusal of both slavery and freedom as founded within the scene of chattel slavery. She hears within black nonperformance as "improvisation" that which cannot be (fore)seen, or registered, by politics, by sovereignty, and the state.[8] This unforeseeableness, which is to me also an unregistrableness, because it moves in all temporal directions from future to past, and also to the present, is a disturbance within contract law that precedes even its beginning. This case bespeaks an impossible sight, where the white gaze cannot see, where blackness plays out life outside of individuals with a standing in politics on the basis of the contract as measurable relations.

What if the *return*, which can be expressed also as a staying put, of those Black Swiss expelled from Switzerland as a space of citizenship is, like *Betty's Case*, a transmutative act that respells citizenship itself along the lines of an impossibility? What if living as Black, while there should be none, is a nonperformance of the terms of being (Swiss or citizenship)? Can we belong without belonging? Can belonging be removed from a measurement based in the delimitation of a proper life, of a proper kinship? Can we make relations in excess of those who should exist? What is the status of citizenship in Betty's case? Just like *homo sacer*, who presumably remained a citizen, there is a kind of ambiguity to Betty's status. How can homo sacer be a citizen without the right to the inner space of citizenship, which was in his time the city (the word *citizen* derives from *civitas*, which means "city")? There is a kind of strange unaccountability, uncertainty we might have to say, of the positioning to citizenship that these two cases bespeak. What is this kind of (non) citizenship that exceeds both citizenship and noncitizenship? While there are today many intermediary steps to citizenship, especially in Switzerland, with its laws of lower-level passports, *sans-papier* (without-papers), refugees, and other kinds of immigration statuses that can last a very long time and often are a kind of limbo, I don't mean, with this uncertain space, these kinds of controlled spaces that are mainly created to exactly exert control over this kind of interstitial positioning.[9] These categories function, like the category of nonperformance, as a cover-up of a lack of a system of sovereign states to be all-encompassing; these terms/names entail physical attempts toward controlling and correcting this failure. These nonexistent spaces trouble a world where everyone must belong clearly, both through papers and through other ways, such as biologically, to an authorized sovereign space. This interstitial space, that is called forth by nonperformance in *Betty's Case*, seems to bespeak an ambiguity of status that requires the refiguring of citizenship, of space, of the world, itself.

Black kinship was breached by property law because property law demands the splitting-up of the Negro slave family. Thus, Black lives practiced alternate forms of kinship.[10] In *Betty's Case*, it is legal *freedom* that would break up these Black kinship practices. Fred Moten explains:

> The invasive property relation is a mere example of a general condition of slavery as contract, which must include all variable obscene forms of unfulfilled promises. Notably, though Spillers references specifically the "property relation" as that which interferes with otherwise established (however precarious) legal relations of black kinship, in *Betty's Case*, it is not the property relation, but the legal freedom of contract, that invades.
>
> Second, and more important for my discussion, this "enforced state of breach" is itself a reference to contract, but by promises that are known only by their inherent brokenness. The temporality of the promise represented by Spillers's phrase, "enforced state of breach," here is not of future satisfaction or fulfillment foreclosed, but a condition in which compromise with any future called forth by the promise of human kinship is impossible (which is not to say that promises of some

other kind of relation are not forged). We must retain this nuance. On my reading of it, Spillers is not arguing that the law of property denies the slave the capacity to promise, for some form of contractual relation (kinship) must exist in order to be "*invaded.*" Instead, she is arguing the reality of a "state of breach," which through its phrasing, implicitly introduces the need for a specific understanding of the nature of the promise, on the side of the slave, in a context where both legal unfreedom *and* freedom are enforcements against the radically indeterminable future invited by the legal idea of nonperformance. Thus, as she states, the contractual relation of "'kinship' loses meaning," which is to say neither that whatever Betty might have promised no longer exists, nor that the reality of unkeepable promises forecloses the radical undeterminable futurity contained in those promises. Rather, we are on the terrain of a relationality based in an unnotarizable promissory note, already breached because unfulfillable. The performative of blackness as nonperformance opens up onto a form of intimacy that registers only in an obscene form of consent to being bound through such promises.[11]

This alternate kinship practice both refuses the control of bloodlines as markers of proper biological belonging, which is tied up, in the case of Negro slaves, in belonging as property to a master of a house, which is to say that it is also a *geographical* question. What is announced in Black kinship is the excess of relations, a form of belonging that cannot be accounted for, new kinds of connections that break over borders of proper ways of coming in touch.

If homo sacer, as exception that proves the rule, instantiates the authority of the sovereign, then thinking with Blackness and its excessive relations is to shift this scene from exception as a negative to be held in opposition to where we flow with this hole in it all, a hole that bespeaks more than all there is.[12] Black kinship is that which *breaks* kinship itself, breaks it up, like a breakbeat, into other relations.[13] It *shifts* what kinship means because of what blackness does to kinship. In this same sense, since they're entangled, if Blacks have citizenship in Switzerland, then citizenship must also shift its meaning. What kinship, what belonging, and relations in general, *means* becomes changed in this scene of Black motherly practices.

In the United States, where *Betty's Case* took place, citizenship is ascribed through jus soli, by being born there. While this law of citizenship traces back to English commonwealth law, there is a kind of excess that emerges in Moten's theorizing of *Betty's Case* that uncovers a gap that citizenship law itself attempts to continually erase. This erasure happens on the order of an avoidance of the way in which kinship plays into, and is fundamental to, national belonging. Jus soli does so via the erasure of biological, or *familial*, to be more precise, kinship, and jus sanguinis through the erasure of how biological kinship is about geographic spaces and movement through it.[14] This failure is of this distance that is necessary, between any kind of claim or carrying of citizenship, of having the attribute of the citizen, as being a part—that is, also as being the nation itself and the nation. In other words, the nation and the citizen must at once be the same and not the same thing, for the citizen is the nation and the nation is its citizens. Between jus soli and jus sanguinis, as their denial of each other in each other, is an attempt to cover over this gap that is fundamental to citizenship.

Since the beginning of modern nation-states, Black people have been raising petitions and making legal claims that used the very definitions of citizenship and freedom to fight for their own—not as a mere matter of assimilation but, rather, more radically, it is the very refiguring, the rewriting, of freedom and citizenship.[15] For example, we may think about this gap within citizenship as a gap also within kinship, and questions of belonging, and/or relations, in general, in relation to the Fourteenth Amendment, which gave citizenship to any person born in the United States. The need to instigate a change in citizenship laws reveals a moment of breach that citizenship itself covers over in its claim to properness, in its distancing from its borders. *Betty's Case*, which took place before both the Thirteenth Amendment, the abolition of slavery, and the Fourteenth Amendment, poses this very problem, the problem of citizenship's lack of proper authority over all life, claimed in its authorized control and identification of its borders. It seems that the Fourteenth Amendment at once reveals a radical refiguring of citizenship while also marking a kind of reinstating of control. If the 1857 case *Dred Scott v. Sandford* reinstated control over slaves who claimed freedom because of having traveled from the South to the North where slavery was abolished, then it was the Thirteenth Amendment (1865) that countered this reclaiming of control, a control that was exercised over the kind of nonperformance that Betty poses us to think. In other words, *Dred Scott v. Sandford* seems to have been a remedy for the failure to account for a kind of excess of accountable belonging, which the Thirteenth Amendment both countered but which also seems to be, in some ways, continued in it. The continuation of the kind of control over this undefined space is exemplified by black codes, which were used in the South to regain control over Black lives after the passing of the Thirteenth Amendment. Freedom implied also their gaining of a political and societal voice, a freedom, or a definition of freedom, which was also vehemently attempted to be owned and protected against Black lives. This is exemplified by the growing resistance to Black voters and their political agency during the Reconstruction period after the passing of the Fourteenth Amendment and the Civil War. The efforts toward giving African Americans the right to vote and to affect the political structure of the country, as historian Tyler Stovall explains, "were fought tooth and nail not only by white Southerners and the Democratic Party that represented them but also by Andrew Johnson and other moderate Republicans. Resistance to Reconstruction frequently turned violent, especially after a group of ex-Confederate soldiers founded the Ku Klux Klan in 1866, unleashing a wave of white terror against Black and white Republicans throughout the South."[16]

In this sense, Black lives become excluded from political agency, even in freedom. Thus, the question of being able to vote is also about the question of what a country (and in this sense, also freedom) is. Voting becomes a site for protest as well as a symbol, and actual way, for the definition of not only what it means to be a citizen but also to be *free*.[17]

While there seems to be a series of "nonperformances" within these shifts in citizenship, there is also a countermovement of regaining control over Black lives, or the nation's borders in general. There is a clear closeness between being controlled and being on the border zone of citizenship as an illegal, or legal, immigrant, as a Black person whose rights to citizenship is continually put under scrutiny and denied through a whole system that opposes Black lives, and a nation that attempts to own and control even its border zones. A telling example of a resistance to both slavery with citizenship, as well as a reclaiming of the definition of such freedom, is from Phillis Wheatly:

> Wheatly accompanied her master Nathaniel Wheatley to London in 1773, less than a year after Mansfield's decision [which declared slavery illegal in England]. Although she did no more than move between the colony and the metropole, her political status altered significantly: in Boston she was a slave; in London she was not a slave but a subject. She likely recognized the opportunity the transatlantic trip presented, for on 22 September 1772, the *Boston Gazette* reported events in London relayed by a correspondent who had observed "that as Blacks are free now in this country, Gentlemen will not be so fond bringing them here as they used to be, it being computed there are now about 14,000 blacks in this country." Wheatley might have stayed in England, but instead, from her new location and status, she negotiated her manumission, a contract she astutely had bonded in London, thereby guaranteeing that she would return to Boston a free woman.[18]

But rather than focusing on the ways in which control and power are reclaimed, I want to think about the steady insurgence work behind the scenes that often gets subsumed by a narrative that this claim to authority itself continually requires. And it is this underground that is the excess of the gap between citizen and citizenship, between subjects and a nation, that we must think from and think about with, and for, Black lives.

Politics of the Womb

In the Swiss context, as uncovered in chapter 2, the control of the border, and thus the nation, is revealed to be deeply entangled with the managing of reproduction as a question of citizenship and kinship formation. To do so, there is the need for a continual identification, or articulation, of failed kinship. The emergence of proper belonging as a question of the right to enter contractual relations, and to be in proper kinship relations, requires the control over those who are outside of it. Those who are outside of relations are controlled by way of how they may join the inside. This means manumission but also, or today, the *path* to citizenship. Such requires the initial identification as danger and the subsequent control over such danger by way of identification, both happening at the same time. Whether the result is the denial or inclusion depends on the ability to leave blackness behind, which is also to say that such is circumscribed by the delimitation of such excess through it being held, absolutely, in black. This question of relation, as a question of kinship, is thus the erasure of the "biological" of the citizen, as the erasure of the possibility that the citizen may radically change.

To move by way of the interstitial—that is, in excess of borders as a measure of control, as borders that, through moving across them, become a technology toward rewriting our stories, our lives, like Wheatly's crossing over so as to then return—I engage a force that may turn relations over into music. In the underground of relations as border zone of kinship and belonging is motherness, or motherly practices, or the mothership. Scholar Hortense Spillers elaborates Blackness's law of the mother:

> The African-American male has been touched, therefore, by the *mother, handed* by her in ways that he cannot escape, and in ways that the white American male is allowed to temporize by a fatherly reprieve. This human and historic development—the text that has been inscribed on the benighted heart of the continent—takes us to the center of an inexorable difference in the depths of American women's community: the

African-American woman, the mother, the daughter, becomes historically the powerful and shadowy evocation of a cultural synthesis long evaporated—the law of the Mother—only and precisely because legal enslavement removed the African-American male not so much from sight as from *mimetic* view as a partner in the prevailing social fiction of the Father's name, the Father's law.[19]

The touch by motherness bespeaks a making of relations in excess of delimitations, or prefixed circumscriptions, of what relation means, which is also to say, how and with whom relations can be shared. After Spillers's work, there is thus given, in the example of Black mother's law, a radical refiguring of relations. As a question of kinship and relations articulated by what it means to mother, and how motherly practices can be practiced (not only by women with children but by all who are touched by motherness), bespeaks also citizenship, itself derived from kinship and relation to land.[20] There is a radical incursion into citizenship and its mechanisms of power that we must make here, a critique that subsequently turns into an outer-space flight with the mothership. Firstly, citizenship is tied up in the commodification of women's reproductivity. Secondly, Blackness asks us to think our relations in excess of control over the boundaries of motherness and relations.

Seminal French scholar Luce Irigaray poignantly analyzes in *This Sex Which Is Not One*:

> The passage into the social order, into the symbolic order, into order as such, is assured by the fact that men, or groups of men, circulate women among themselves, according to a rule known as the incest taboo.
>
> Whatever familial form this prohibition may take in a given state of society, its signification has a much broader impact. It assures the foundation of the economic, social, and cultural order that has been ours for centuries.
>
> Why exchange women? Because they are "scarce [commodities] . . . essential to the life of the group," the anthropologist tells us. Why this characteristic of scarcity, given the biological equilibrium between male and female births? Because the "deep polygamous tendency, which exists among all men, always makes the number of available women seem insufficient. Let us add that, even if there were as many women as men, these women would not all be equally desirable . . . and that, by definition . . . , the most desirable women must form a minority."[21]

In Irigaray's analysis, women are commodities because reproduction must be controlled for and by the nation, and the exchange of women is the foundation of relations among men.[22] Since the *accumulation* of things, of commodities, rather than their inherent usefulness, constitutes wealth,[23] women, as abstract commodities, are counted not in their usefulness but in how many man or society possesses.[24] Irigaray furthermore elaborates that it isn't "properties" ascribed to women's bodies that thus make them of value; but rather,

> her body constitutes the *material* support of that price. But when women are exchanged, woman's body must be treated as an *abstraction*. The exchange operation cannot take place in terms of some intrinsic, immanent value of the commodity. It can only come about when two objects—two women—are in a relation of equality with a third term that is neither the one nor the other. It is thus not as "women" that they are exchanged, but as women reduced to some common feature—their current price in gold, or phalluses—and of which they would represent a plus or minus quantity. Not a plus or a minus of feminine qualities, obviously. Since these qualities are abandoned in the long run to the needs of the consumer, *woman has value on the market by virtue of one single quality: that of being a product of man's "labor."*[25]

I would claim that these women do have something like property; namely, that they exist as proper life, as members of society—they are *white* women.[26] In Irigaray's elaboration, what comes to the fore is a certain closeness of women to blackness, on the order of them being like or close to a border zone of the species. In other words, there is a certain closeness, after Irigaray's elaboration, between women and the owning of Black lives as slaves—commodification of life.

There is an entanglement between jus sanguinis as a way to control who may become a citizen, this analytic of women as commodity, and antiblackness as a structuring paradigm of modern subjectivity/citizenship.[27] For, it is the forming of a border zone of the species by way of the control over reproduction, and thus women, who symbolize reproduction.[28] In this sense, blackness is a repository of reproductive vitality that must be both used and absolutely separated from the world of man.[29] This is why white women were seen as close to black. They have to be close so as to have reproductive capacity. Noémie Ndiaye, for example, shows how seventeenth-century Europe minstrelsy involved a performance of erotics that was geared toward white women. Two points are important here. Firstly, these minstrel performances mainly staged erotic relations between white women and black (that is, blackface) characters but not black women and white men. Secondly, it was black men who both are overly sexualized and ultimately saved, and/or made whiter, by way of proper love.[30] Thus, there is a kind of *nearness* that the (white) *woman* must have to blackness because she must be able to bear children while having to be distant enough to not be able to become black.

Antiblackness requires the control of sexual reproduction to protect itself from blackness. That is to say that both the violence against women and the control over reproductive capacity of a nation (or whiteness) are about controlling blackness—which in the same turn comes to mark the border of the proper species/subjects. This is why the control of gender in Black lives is crucial, for if the claimed excessive sexuality of Blacks is not controlled, then whiteness loses its value. Thus, antiblackness has to at once make blackness into the symbol of sexual reproduction itself and at the same time own it, claim ownership over life and its production or proliferation.

In Irigaray's analysis, civil society based on contractual relations is rooted in the exchange of commodities, is in the business of a politics over the womb, over motherness. The importance of "women as commodities" lies in their providing of "natality"—being able to provide *birth* rates that are measurable, as protection and continuation of the species—continuation also means protection from the corruption of the species by bad or impure blood.[31] Women are to be controlled as sites of possible miscegenation, which is of a closeness to blackness that would be less than proper life. But Black women's reproduction was during slavery both a source of future profit as well as a negative for the value of that which brings profit (i.e., slave labor).[32] As Iyko Day argues: "Black women's labor is associated with a deviant form of value that poses a quantitative threat to the relations of white social reproduction."[33] Since Black reproduction is associated with both having more Black labor and also the devaluation of each individual slave, or labor unit, Black women's wombs are a site of danger. Black women's wombs are being controlled so that there is only the right amount of Black labor, so that value does not become negative. Black women's reproduction is thus both possible devaluation but also the source of that which carries value (i.e., the owning of slaves). Day turns this into a radical resistance made possible through thinking with Black motherness, Black women as what continually undoes the very definitions of value defined *against* Black lives whenever they claim decisions over their own bodies and reproduction. Her analysis shifts our understanding of Blackness from being a proper abstraction to, what I might term, *a disruptive abstraction*, which we could also say is an overflow of the interstitial between abstraction and sign, turning deviant reproductive labor of Black women into a performance that undoes and rewrites value.[34] In my reading, this is also a radical refiguring of belonging as a question of citizenship, or relations in general, that is announced in Black kinship's foregrounding of mother's law.

If women are controlled to protect the species, to protect relations, then in such ownership of women is expressed a delimitation of life, of the biological as well as the social, as an extension of biopolitical and necropolitical enforcement of protection and formation of a proper species as a proper life. It is here that a certain register of antiblackness becomes apparent, one that also requires the commodification of women, that at the same time creates a distance between women who are not Black and Black women (and those closer to blackness) as a question of natal politics.

Rosalyn Diprose and Ewa Plonowska Ziarek elaborate on the relationship between biopolitics and the need to control life as a concern of controlling women's reproductive rights. They reveal how biopolitics is in the right to let die but also to make life.[35] The control of birth rates is a question of protecting citizenship:

> Since the French Declaration of the Rights of Man in 1789, he [Michel Foucault] argues, it has been "simple birth" (which he equates with "bare life"), rather than the "free and conscious political subject," that has been "invested with the principle of sovereignty" in the "passage from subject to citizen." In this way "birth" has become equivalent to "nation" as one (fragile) basis upon which "man" is afforded rights. Agamben is of course wary of this biopolitical unification of the "principle of nativity" and national sovereignty. Significantly, though, as a consequence of ignoring Arendt's idea that the meaning of the event of human birth plays a central role in shoring up the political event of natality, Agamben makes no mention of what should be obvious: *that women and maternity must be key targets of biopolitical erasure of political natality* (agency) in order for this reduction of "birth" to "bare life" to "nation" to occur.[36]

Basing their work largely on Arendt and her elaboration of natality, the authors go on to a series of brilliant revealings of modern nation-states' investment in controlling women's reproduction, or natality, which they think in Arendt's sense as the basic capacity for newness, including but going beyond biological birthing.[37]

Not only is power exerted as a means to force birth in the form of, for example, antiabortion laws but also as a form to control birth through sterilization and contraceptives. In their analysis of Aboriginal women in Australia, as well as Indigenous people in the United States, Diprose and Ziarek uncover the entanglement between racism, nation-building, and biopolitics as a form of natal politics through limiting access to abortion pills to sterilization.[38] We can expand this with Bonaventure Soh Bejeng Ndikung's brilliant work *The Delusion of Care*,[39] to contraceptive policies and discourses where Black women's reproductive "agency" is controlled by institutions with investment in protecting society from Black over-reproduction, a reproductive capacity that is perceived by the world, as François Vergès elaborates, as a "terrorist threat."[40]

The control over women's reproduction thus emerges out of antiblackness: antiblackness must control what constitutes life on the basis of exercising authoritative control over the borders of life. Exemplary again is Ndiaye's analysis of seventeenth-century Europe. Ndiaye points out how in Renaissance performances, particularly in France, Black women were willfully erased: "Their absence results from compulsive acts of erasure triggered by anxieties regarding their all-too-well-known desirability. At the root of Afro-diasporic women's exclusion from the interracial erotic economy of the high baroque stage in France, we find shame and denial about colonial realities."[41] Thus, the erasure of Black women was the erasure of the sexual desire of white men and the sexual violence enacted in the colonial context against Black women. This both removes sexual desire from man and reenforces the border to blackness. Thus, at once white men are without sexual deviancy, marking blackness as sexually excessive and white men as sexual norm, while also creating an absolute separation between whiteness and blackness.

In fact, these kinds of tropes of excessive reproductive capacity, and blackness as repository of sexual libido, can be found even today. Exemplary are the discourses around Henrietta Lacks's stem cells, which have been the foundation of much of modern cell research and were taken without ever getting her consent. Her cells often have been described as having an excessive *reproductive* power. As Jayna Brown elaborates:

> We see this anxiety around HeLa [Henrietta Lacks's] cells, as they have spread rapidly and thoroughly. Because of their initial wide dispersal, HeLa cells reside in laboratories across the nation. The cells' "unusually malignant behavior" has enabled them to invade all cell lines that scientists believed to be separate, invalidating decades of cancer research. HeLa's legacy of sabotage lives on, with the 2013 discovery of HeLa contamination found in bladder cancer cell lines. The initial discovery of the contamination instantly brought to the attention of scientists and

the public the fact that Henrietta Lacks was black, and the first scientist to declare the contamination assumed that her race was the reason for the cells' unusual power to reproduce. All sorts of racial narratives resonated with this case, including white fear of racial miscegenation and the oversexualization of black women, as well as the injustice of Lacks's family surviving in poverty while huge profits are made from their mother's cells.[42]

Natalpolitics thus bespeaks how there is not simply the control of life with death but also with life, or more precisely reproduction as reducible to birth as a singular event, as border zone to the (racialized) species. As Ruth A. Miller elaborates:

> Rather than understanding men as the norm and women as artificial facsimiles of men, it makes far more sense in a biopolitical framework to understand women as the norm and men as their copies. It is the womb that has become the predominant biopolitical space, it is women's bodily borders that have been displaced onto national ones, [and] it is thus the citizen with the womb who has become the political neutral—and rather than grudgingly granting women the artificial phalluses assumed by liberal theory, one can in fact advance an argument that men instead have been granted the artificial wombs assumed by its biopolitical counterpart.[43]

We have to read this elaboration after Irigaray and after our elaboration of a certain assumed goodness of civil society and the signifier of the human species. It becomes clear after Irigaray that such womb, although of the woman, is how society owns women, as parts to protect society and the species. This species as society is proper life and thus nonwhite, Black, and women in/close-to black become antagonistic to the survival of the species.

Catherine Waldby and Melinda Cooper's essay further evinces how women's reproductivity is not about women's agency but about exchange relations, and such becomes evinced analytically with an attention to Black women and women close to blackness: "New reproductive technologies like IVF have disaggregated it from its *in vivo* location, and stem cell technologies have diverted it into biomedical domains unconcerned with the production of children. Reproductive potential is now bifurcated. *In vitro* embryos and *in vitro* oöcytes can be transplanted to produce another human life, a child; and they can be biotechnically reconfigured in a laboratory, diverting their pluripotency into the production of embryonic stem cell lines."[44] Thus natalpolitics owns women's parts in the form of national blood and mother tongues, as well as women's wombs or motherness in general.

While writing this book, the United States, for example, just reinstated more restrictions on women's reproductive rights with the overturning of *Roe v. Wade* in 2022. People who have spoken up against it include Michele Goodwin, whose scholarly work points out how the control of women's reproductive rights in the context of the United States has been a question of antiblackness. In her response, published by the *New York Times*, she points out how "ending the forced sexual and reproductive servitude of Black girls and women was a critical part of the passage of the 13th and 14th Amendments."[45] In her book *Policing the Womb*, she provides a thorough analysis of the myriad of ways in which women's wombs are being "policed" by the state and how such intersects with antiblack paradigms. As she states: "The book illumes how the recent robust lawmaking that restricts when and under what circumstances women may access reproductive healthcare rights functions not only to undermine women's constitutional rights but also leads to the surveillance of their reproduction, criminalization of their conduct during pregnancy, and ultimately the burdening of their health."[46]

It is clear that the control of reproduction bespeaks both a violence against women as well as the delimitation of who may be born, belong to a nation through kinship, and is thus a nexus of both racialization and national belonging. Diprose and Ziarek in their account of politics over natality in relation to race also elaborate that "the biopolitical regulation of reproduction collapses the traditional political distinctions between private and public, the biological and the political."[47] This problematic of the collapse between the "political" and the biological, where politics is about preserving one kind of biological, as I read the authors to be saying, they propose can be countered by extending the access to the political. They emphasize:

> Moreover, the plurality of human existence is not just in terms of a multiplicity of distinct entities (it is not an individualistic notion characteristic of neo-liberal pluralism). Rather, natality is witnessed, "welcomed" and "disclosed" to and by others as a "who" (a beginner whose "personal identity" is undeterminable, or "unique distinctness" per se) as opposed to a "what" (a set of particular characteristics). This idea of the disclosure of natality *to and by* others forms the basis of Arendt's unique notion of the political, which is not the realm of representative government but the public "space of appearance" consisting of the inter-relational disclosure of natality through speech and action. The

disclosure of natality or uniqueness within the togetherness of human affairs is also the basis of Arendt's notion of what Kant calls human "dignity" and the principle underlying human inter-relationality, democratic plurality, and ethical community.[48]

There is a radical possibility in this public as a togetherness that is foreclosed by natality's requirement for absolute control over births. At the same time, we must think critically about the notion of how natality comes to the rescue against the collapse between politics and biology, or how "the plurality of human existence" is supposedly not a "neo-liberal pluralism." How do we come to grips with the fact that it is exactly on the basis of a plurality and diversity in and of the human that antiblackness becomes articulated? They presuppose, and inherit from Arendt, the idea of a pregiven space that already delimits and bestows upon those who are a part of it the ability to enter the political, or citizenship, and the human. Seyla Benhabib's critique of Arendt is helpful here:

> This anthropological universalism contains an ethics of radical intersubjectivity, which is based on the fundamental insight that all social life and moral relations to others begin with the decentering of primary narcissism. Whereas *mortality* is the condition that leads the self to withdraw from the world into a fundamental concern with a fate that can only be its own, natality is the condition through which we immerse ourselves into the world, at first through the good will and solidarity of those who nurture us and subsequently through our own deeds and words. Yet insight into the condition of *natality*, while it enables the de-centering of the subject by revealing our fundamental dependence on others, is not adequate to lead to an attitude of moral *respect* among equals. The condition of natality involves inequality and hierarchies of dependence. By contrast, Arendt describes mutual respect as "a kind of 'friendship' without intimacy and without closeness; it is a regard for the person from the distance which the space of the world puts between us." It is the step leading from the constituents of a philosophical anthropology (natality, worldliness, plurality, and forms of human activity) to this attitude of respect for the other that is missing in Arendt's thought. Her anthropological universalism does not so much justify this attitude of respect as it presupposes it. For, in treating one another as members of the same species, we are in some sense already granting each other recognition as moral equals.[49]

How might we come to think in excess of natality as a singular event that presupposes being allowed in, into society, into the public (that also relies on the exclusion of others)? Might there be a way to refigure our appearance, as our entrance into society, this scene of natality, and move by way of those whose denied entrance is always already the backdrop of natality? In other words, if our death, life, and birth are all borders to society, to the world of proper citizens, then it is in the excess that an entrance always already requires, and antiblackness aims to control/police, that we can hear the possibility of our coming to matter.

Getting in Touch

Moten's critical engagement of Arendt, and her notion of political agency, which is rooted for her in this event of natality, in being born into citizenship, can bring us to thinking alongside this underground:

> What if the (under) commonality movement, which proceeds outside of normative political agency but is also illegible to and disruptive of already existing society in its constant regulation of sociality on behalf of the political, was not about sacrifice or accession but about revolution, if by revolution part of what is meant is the ongoing refusal of the artificial boundary between sociality and publicness whose maintenance was, for Arendt, the politicotheoretical justification for school segregation. This is to say that corollary to insistence on a variation of Arendt's distinction between the social and the political is the rupture of the dividing line between private and public.
>
> What was at stake in Little Rock, what was being asserted insofar as it was being defended, was free association (though perhaps the term *fugitive sociality* is more precise), which politics and the family, the public and the private, all are meant to regulate. Free association, in this regard, manifests itself in desegregative planning, not integrationist achievement.[50]

While one could read Moten as advancing a collapse of social and public, or private and public, we mustn't run to this conclusion, as one can also read this passage in a slightly different manner. In other words, what I'd like to propose is to follow Moten's text, which means to follow these fugitive socialities or this free association, that bespeaks *a particular kind of relation to what we might term* a border (such as, for example, natality bespeaks). What if it is not about the collapse of borders, not about a kind of erasure of borders, but rather about how there is a movement at work here, a crossing of borders that undermines the very properness of borders, the authority of the inside to police, control, and own borders (the underground of the name *border*, we might also say). What comes to us, in this scene of sitting with borders, is what *fugitive* always denotes: our sociality as the crossing of borders when breaking into the political through our free association across borders. Our free association across borders is our claim to names that is also our unnaming, our renaming, and our refusal of a proper and its claim to being proper by owning and controlling borders (in the case of *Black*). It is a breaking in—into citizenship, into schools, into universities, into citizenship—across borders and in the process undoing and redefining what borders are. Borders turn into interstitial spaces, out of which we make relations anew, where abstraction and name engage each other, uncomfortably close so that new names must emerge.

Return, as in *nonperformance*, thus bespeaks a kind of breaking of borders that refuses a certain temporality, or narrativization, necessary for the upholding of proper racialized national belonging.[51] In this way, the "fugitive sociality" that is the sharing of Black Lives Matter protests, of blackness/Black in general, across borders of states, bespeaks this. Similarly, the very emergence of the term *Black* cannot be separated from this rupturing of controlled delimitations. Spilling over into the world of proper relations is an uncertainty principle—or we might call it a silence, one like Schrödinger's cat, that exceeds the recognition, or identificatory and disciplinatory mechanisms, of an antiblack world. The definition of life, and with such the racialized delimitation of citizenship, which is also the invention of race as much as it is the control over life that is continually changing (biologically as well as otherwise), rests on the control over the borders of the nation, of the species. But this also means that the borders are where the species, where life, where citizenship, changes. Thus, blackness is also the revealing and changing of *our* colors (or the world); it is the name for the very refiguring of relations, for the very restructuring of the terms (of life). And *relations* here is not limited to kinship but also means logic, as in the structuring principle of grammar, or language, science, and also structures in the world, such as nations or institutions.[52] As relations' excessive precondition, blackness is the changing of change itself, and with such shifts what relations may mean.

This movement in excess of bordering wholes is given to us as thought by Spillers's work, to *identify*, in *free association* beyond borders of relations.[53] It is an abundance of a Black feminist practice. According to Alexis Pauline Gumbs:

The main argument of Lorde's article is that as black women "we can learn to mother ourselves." This statement comes after a section in which Lorde explains that black daughters often believe that no other person will be able to provide them with the love and understanding that they have learned to expect from their mothers. Lorde wants to counter the belief that only black women socialized into a mother/daughter relationship with each other can provide the mothering that healing and community building requires. But it is significant that Lorde does not say "we can learn to mother each other." She says instead "we can learn to mother ourselves" which relies on an intersubjective production of a rival maternity, that does not reproduce familial relations, but rather disperses the labor of mothering.[54]

In the underground of borders and policing as the instantiation of borders, as places to claim authority over life through delimiting life, limiting women's reproductive rights, limiting kinship and relations, is a black-hole kind of story, where a motherly law bespeaks an excess of relating. Black lives, as revealed by Black women, are of mother's law, where relations are in excess, are fugitive socialities, across borders, entering into unspeakable relations where relationships become outer-space flight. The very term *border* bespeaks all these techniques of policing and control that are antiblack delimitation of life for the claim of an authorized, sovereign, and proper one. But behind borders is the nameless and unnameable that we call *blackness*. The mothership—the free association opened by blackness—is not only the underground of borders but also its before, after, and excess. In thinking with this black nonspace, we may come to think of the way in which singularities bespeaks not a one, an individual, in opposition to, or friend-enemy relation with, other ones, but rather as black holes teach us, singularities are where the definition of space and time itself ends. We think from the empty abstraction behind the name, that is not a lack of a name or the negative of all these things in relation, of value (economics or grammar), but rather names' unending creativity expressed in (un)naming. This hole at the center, which we found through the border, or the in-between, is the not-in-between, the emptiness that exceeds its naming in relation, the unknown that cannot be tamed by an already given world.

7

Interstitial Listenings IV

*8'46" George Floyd in
Memoriam and White Gaze II
Black Square*

[ ]
Interstitial listening is a kind of underground of the in-between—maybe something like taking a breather, a pause or break to breathe. While this might seem like an in-between moment, isn't breathing really always there, not just in between? Doesn't breathing resist a kind of classification by those things breathed? Is breathing maybe not even about breathing, as in reducible to a corporeal use value? Like when breathing becomes a practice to return to before any sense of time that exists always there, outside of the measurement of life with death. A practice of breath is, on the one hand, a way to rest and reset and, on the other, an activation, a raising of the fire within—a protest. We need interstitial listening as these gaps for air that become timeless, this intimate exchange with the wind. But this break, this break-work, a breath-work, like learning to breathe, is here also a fraught scene. The sound of air, of wind, of breath so familiar, is taken from some of us, and particularly from Black lives. But this music of lives in excess of a distinction between legible and illegible, sound and noise, that breathing bespeaks is a radical musical

practice. All these complex things make our world, but breathing remains its interstitial space—one from which we listen, think, and make our sounds into music; a silent noise of resistance is breathing while Black.[1]

Time is the end of breaths and also their opening. Time flows from breath but becomes detached from it, becomes a measure of ended breaths (notice this is a music thing). I turn back time, where time itself comes from breath, where our breathing is our making of time, making time to think about lives. But we begin with an homage to a series of protests against the antiblack closing of breathing that listens to breathing in Black as the very danger of too much breathing, of too much life, that can also be thought of as a kind of breathing beyond air, maybe underwater, to hint at the rumination with Black ancestors in the waters, there in the Atlantic despite cutting off their air. This is why we breathe. Black lives breathe for each other, in every breath they take.

8'46" George Floyd in Memoriam

Eight minutes and forty-six seconds, or 8'46", is the duration that marks our breathing in concert for those whose breath has been brutally stolen, taken by an antiblack system. Time becomes our protest in silence and also a celebration, a celebration of the life that is George Floyd, that is Mike Ben Peter. Eight minutes and forty-six seconds became a symbol for protests in 2020, after George Floyd's murder, as it was the number initially stated in the charges raised against police officer Derek Chauvin, who kneeled on Floyd for this amount of time, by the prosecution.[2] This number was later revised to seven minutes and forty-six seconds,[3] and only after the release of the body-cam footage was the number corrected to nine minutes and thirty seconds.[4] Charles Uzor's first piece as an homage to Floyd, written in 2020, is titled with this duration: *8'46" George Floyd in Memoriam*. Written for "any number of musicians with any instruments," it consists of seven minutes and forty-six seconds of breathing and one minute of silence.[5] The duration—here musical time and form—becomes a measurement of antiblack violence as well as protest against it.

The work also references, through its similarity of form, content, and title, the groundbreaking work *4'33"* by John Cage, which consists of four minutes and thirty-three seconds of silence. To unpack the critique raised by this work, or what we could also call this incipit for thought, it is helpful to consider the *specific* questions raised by silence in the multiple registers of its appearances and contexts. Specifically, as an appearance of silence, this musical work comes to the fore as a problematization of silence, indicated in this superposition of Cage and Floyd (as well as the silence in the background of the work, around Blackness within Switzerland, and Black composers in general, as well as the silence in the music world in relation to questions of race pre-2020). Fundamentally, the piece is a staging ground for encounters to rethink assumptions of music's relation to the political and the social, to lives lived in general, and those forces that brutally attempt to control and manage lives.

Cage's *4'33"*, four minutes and thirty-three seconds, marks a duration of silence, silence to be listened to, but it also pushes sounds ignored into our ears. It makes us listen to unthought sounds and noises.[6] Its famous score consists of three movements, all absolutely silent. The piece problematizes what sound and silence are. Cage points to how the piece demonstrates that there is no silence and that *accidental* sounds, as in the sounds from outside of the concert hall as well as the audience themselves, always happen during the performance of *4'33"*.[7] The valence is placed upon sounds that are without any cause, especially the composer's own intent—this absence of intent was of fundamental importance to Cage as a composer. In some ways, Cage is pointing to the audience as well as the world as music. I would say that Uzor's reference engages this refiguring of listening.

But, at the same time, a critique must be raised in relation to Cage's insistence on the absence of intent. Limits of what *no-intent* means for Cage can be heard in black. I'm thinking of the example of Julius Eastman's performance of Cage's *Song Books*, which was too far away from Cage's own nonintent for performance. When discussing his issues with the interpretation of this particular performance, an anecdote of Black lives out of bounds comes to the fore.[8] The limits of chance—that is, the absolute absence of intent as musical composition—are colored black.[9] What escapes Cage in his insistence on equally valued sounds without bounds in *4'33"* is noise in black, is black noise, is exactly what the piece wants to thematize: silence. Silence becomes sounds and thus silence is erased—which also reifies silence as the negativity, the absence, of sound. But *4'33"*, as *8'46" George Floyd in Memoriam* teaches us, also speaks another tongue: it can be silence's unboundedness flowing over those holds imposed by duration and place. Cage, in thinking sounds outside the frame of the concert hall—and which can be thought of as outside the walls but also as inside the body; inside the body because in silence we may have the illusion of hearing the sound of our organs like he did during a visit to an anechoic chamber[10]—also *lifts* the cone, that which amplifies sound vis-à-vis silence, that silences silence.[11]

The piece *8'46" George Floyd in Memoriam* as refiguring, or critique, of *4'33"* prompts listeners to rethink silence as not only the absence of sound but, furthermore, as already there in any sound. Silence is heard not simply as not-silence, as (absolute) presence of sounds (without intent), nor as the absence of sounds. Silence is making sounds (in its not-making sounds)—silence is sufficient in itself only if silence is never itself, or in itself. In other words, within the notion that there is no silence lies inherently an idea of silence as absolute absence, which is also to say that silence is defined negatively, is defined by its opposite—presence and/as sound.

8'46" George Floyd in Memoriam radically refuses presence's assumed self-containedness; it doubles down on silence. There are two components to the sound of silence within *8'46" George Floyd in Memoriam*; or, we could also say there are two appearances of silence (both of which then form another kind of silence that is the whole piece). One is the sound of breathing—something that we could hear under a heading of *white noise*. The other is silence, or what we could hear under the heading of *black noise*. Out of these two terms, white noise is more common in sound studies. White noise is listed in the *Merriam-Webster Dictionary* (with its first-known use dated 1943), but black noise is not.[12] Black noise is commonly used in telecommunications—it is listed as an industry standard in the glossary of the Alliance for Telecommunications Industry Solutions (ATIS).[13] *White noise*, as a term, was initially borrowed from the visual domain, usually used to indicate a sound where all frequencies are equally *present*. That means that every frequency audible to humans, from zero to twenty thousand hertz, is sounding with the same energy. This definition is the result of a common misconception about white light—namely, that white light has all wavelengths present at equal power. But, in fact, white light can have many different varying wavelengths and must not even be composed of all colors. *Black noise* is in common usage despite not being officially recognized as a term and is usually meant to mean the *absence* of all frequencies or wavelengths—silence.

Let me further expand our repertory of terms by introducing two further notions from telecommunications as an aid to my analysis. These two terms are simply modifications of the terms *white noise* and *black noise*. One is *noisy black*: black images with noise, perceivable as nonuniformity or white spots. And the other *noisy whites*: white images with noise, perceivable as nonuniformity or black spots.

Thinking through *8'46" George Floyd in Memoriam* with the aid of these terms, the sound of breathing resembles something more akin to white noise, and the sound of silence resembles black noise while, at the same time, the sound of breathing is always noisy—there is no *proper* white noise produceable by nonelectronic means. Black noise is also noisy, but here not because of the presence of white noise, or white spots, but because of the presence of differentiation within the sound of silence itself, which turns it into black noise in the first place. Noisy black presupposes noisy white just as black noise is based only upon the prescription/description of white noise. This does not mean that blackness does not exist but rather that in this scene black is defined as the negative of white: black is figured as the abject. Black noise is the afterthought of the definition of the term *white noise*. White noise was invented first, followed by other kinds of colors of noise. Black noise is, in its definition, the absence of equally present waves of sound/light; the absence thus of the proper equally present distribution of sound—i.e., white noise.

Here is evinced that it is not an *essential* attribute of black that it is the negative but rather that the instantiation of whiteness, or the white gaze, is based upon antiblackness, or an opposing against blackness, marking blackness as abject. As a consequence, if black noise is the absence of white noise, then white is here defined as not-black. Furthermore, it seems that white noise is not only not-black but is the very antithesis of black noise: the equality of all frequencies as equally present requires the eradication of black noise and noisy blacks, because there can be no absences of sound. This is some kind of metaphor, or mirror image, to civil society and the scene Uzor is bringing us to with his homage to Floyd, where to be not-black requires the erasure of all Black lives from inside proper life. Thus, this is analogous to Frank B. Wilderson III's observation that it isn't about the policing of borders as reducible to the figure of the police, but rather it is about how all who are part of civil society are the antithesis to black, are citizens because they do not "magnetize bullets" or the violence of border policing.[14] It is the very claim to equality among all frequencies that erases the very possibility of another frequency, one that cannot be accounted for by this all-encompassing whole (which is metaphorically, and in our analysis, white noise).

If whiteness is those who are part of civil society, and such civility is made upon the killing (in multiple senses such as social, biological, cultural, and more) of that which is called black, then it is making itself antiblack. To put this in terms of sound and music: white noise is the equal distribution of sounds, and black noise is simply defined in contradistinction to such white noise. But there is another story here: this also means that, as a consequence of being only defined as the negative of white noise, black noise becomes that which must be absolutely unmeasurable. In this unmeasurability, black noise always already exceeds white noise and its negatives. In other words, that which is defined as absence exceeds the boundaries of that which thinks it as its negative other.

Black noise bleeds over and out of the container provided by white sounds (the white gaze) of equal power—white noise, here, limits itself in opposition to blackness, that which is beyond and at its borders. Uzor's *8′46″ George Floyd in Memoriam* clarifies (and shifts) Cage's *4′33″* and teaches us that the bleeding of sound *is* silence, or black noise, and not sound (i.e., silence is not sound; rather, sound is silence). While Cage himself arguably did not want to hear the bleed as bleed but simply as accidental *sounds*, potentially removing the sound of silence itself, we can now hear in *4′33″*, after *8′46″ George Floyd in Memoriam*, the way in which silence is not reducible to the absence/negative of sound. This we can uncover further on two registers of analysis: the first is from within the antiblack world as the silence of the duration, as the silencing of the container. The other, we might say, is from the purview of blackness itself and bespeaks the impossibility to really silence those black noises—sounds of protest and sounds of lives.

While 8′46″, eight minutes and forty-six seconds, marks the silencing of Black life, it also marks the remembrance of a silence unavoidable in an antiblack world. But Uzor's piece does so not through a *re-performance*, nor simply through a revealing of the forms of antiblackness—a form where blackness and silence are defined as that which is not the proper thing and then erased either through killing or erasing silence itself. Rather, what *8′46″ George Floyd in Memoriam* teases out is the blackness that is *attempted* to be made silent; blackness is attempting to be erased by an antiblack world. But here emerges a revealing of the blackness within white noise, a revealing of the silence within sound, of how all sound is silence: both silence and sound are noisy. The order of white noise, of antiblack brutality, that which defines itself upon sovereign proper subjects that are equal to each other and rely for such existence on the eradication of blackness (or black noise, or silence, or life), is troubled because that which *haunts* it, silence as blackness, is found within it (whites get noisy *with* black noise). Silence is brought back into the center of this meditation on silence through the doubling of silence, from white noise to black noise. But not only that, white noise is also made again more silent; it becomes noisy whites, white noise with silences in it. And furthermore, through the very way in which this double marks a moment of silence, *8′46″ George Floyd in Memoriam* is about silence; it brings black noise, sounds of protests, into *4′33″*.

8′46″ George Floyd in Memoriam is full of silence and full of noise. Silence is already noisy, white noise is made noisy (i.e., silences are found within it), and thus it is a piece of, about, and for Black lives. Breathing, which might be considered white noise, is noisy, has silences in it, and is of silence. Silence, which might be considered black noise, is noisy, has sounds in it, and is of silence. And at the end, the whole work, that work of silence in common, is music, which is definitely noisy blackness. Blackness is that which comes to us then as the complexity of lives lived—breathing silent noises. It spills over frames, distorts durations, and troubles the categorical distinctions of sound and silence. Breathing under the heading of silence is breathing in black, noisiness in a world of measurements based in sovereign equality. Blackness is that which moves beyond the hold in the fugitive movement that is music.

The silence in *8'46" George Floyd in Memoriam* could be heard as silence, as the end of breathing, as the end of noisy whites and noisy blacks: resolved into white noise and that which it makes itself from in contradistinction—silence as mere lack (which is also to hear it simply as announcement of another white gaze/noise, as sound by and in itself).[15] But such is refused by Uzor's insistence on giving us more silence than silence. Thus this piece is a silence *for* noisy blacks, for those silences that refuse to keep silent, that in their silence remain here and heard, like all those Black lives taken but still protesting, singing, and teaching loudly.[16] *8'46" George Floyd in Memoriam* is a protest piece, while it is, at the same time, a celebration, a celebration of Black lives, a protest for Black lives, a call to action to be with Black lives, to think alongside life as Black.

White Gaze II

There is another photo (visual) to phono (sound) relationship raised in this piece. *8'46" George Floyd in Memoriam* plays into a minimalist tradition of visual art, of painting squares in one color. Particularly, of course, white squares and black squares through Uzor's use of white noise. There is of course also the black square, the posting of a black square on social media that became viral in 2020 and became part of the global Black Lives Matter protests.[17] It is in Uzor's brilliant use of a musical analog to, albeit also a critique of, the black square in visual art that we hear another elaboration of what black squares can do. While there are many black squares in visual art, Uzor's piece seems to me closest to what Akosua Viktoria Adu-Sanyah's *White Gaze II Black Square* does or portrays. This piece is composed of a black square, but it is glossy and reflective (see figure 7.1). It is noisy, like Uzor's *8'46" George Floyd in Memoriam*. We see it as black noise, and as noisy whites, as light that is reflected, it doubles the gaze of the observer.

Figure 7.1. Akosua Viktoria Adu-Sanyah's *White Gaze II Black Square*.

In visual art, the epitome of the black square is Kazimir Malevich's *Black Square*, painted in 1915 (see figure 7.2). Considered usually as a work that exemplifies emptiness, sometimes read as an absolute absence of things, there is an underground story that comes to the fore through Adu-Sanyah's work, through extra visions and a kind of excess of the claimed flatness of blackness on the canvas.[18] X-rayed in 2015, it was revealed that underneath the *Black Square* are colors and a reference to the way in which blackness bespeaks lives (see figure 7.3). As Hannah Black elaborates:

> An extraordinary recent news story reported that X-ray analysis of Kazimir Malevich's 1915 painting *Cherniy kvadrat* (*Black Square*) had revealed a handwritten annotation on the white frame that surrounds the black square. The inscription appears to refer to a racist joke by

Figure 7.2. Kasimir Malevich's *Black Square*.

writer and humourist Alphonse Allais, who in 1897 captioned a black monochrome "Combat de nègres dans une cave pendant la nuit" ("Negroes fighting in a cave at night"). The art historical rupture of the *Black Square*, this radical gesture, turns out to rest, like so much of the history of modernism, on the illegibility of blackness. The painting masquerades as the negation of representation, but in light of the joke about darkness, negation itself becomes representation; what is represented is the nothingness of certain subjects, which indicates a certain nothingness in subjectivity itself. Malevich wrote: "In the year 1913, trying desperately to free art from the dead weight of the real world, I took

Figure 7.3. Kasimir Malevich's *Black Square* x-rayed.

refuge in the form of the square." This break for freedom (for art), this place of refuge (for the artist), is founded on or overlaid on top of black invisibility, itself unfree.[19]

Black hears thus a story hidden to the viewer who sees the square as absolute absence, through her thinking with Black lives, and the kind of extra gazing—abstract art, epitomized in the black square, as a possibility to think about Blackness's radical incursion into the concept of what it means to be alive or a subject.[20]

Subsequent artists followed in the footsteps of Malevich's *Black Square*, such as Cage, who recognized it in retrospect as a precursor, or fellow minimalist artist Ad Reinhardt with his black paintings.[21] Reinhardt's use of the color black becomes a question opened toward thinking with Black lives only through a conversation with the seminal pianist, composer, and poet Cecil Taylor and Fred Moten's writings about such meeting. In other words, for Reinhardt himself, his use of the color black is absolutely separatable from lives and a purely aesthetic object. In the encounter with Taylor, Reinhardt vehemently opposes the idea of black as something other than merely a color—"It's aesthetic. And it has *not* to do with outer space or the colour of skin or the colour of matter" (he actually states that it is the *absence* of color).[22] But for Taylor, black is about lives, about the way in which lives are entangled in each other, about Black lives, about lives in poetic relations.[23] Rather than black being a negative that both subsumes all differentiation, all individuality, into one larger being, or the inverse, of a larger being's attributes distributed among individuals, black emerges as the breakdown of this dichotomy that is rooted in an antiblack gaze. While Malevich's *Black Square* could be read in the same manner as Reinhardt's comments, Hannah Black gives us an alternative reading of black squares (and thus also gives us a new way of seeing Reinhardt's *Abstract Paintings*; see figure 7.4). It is a reading where blackness bespeaks a wholeness that is not about presence, where togetherness results from the holes in it all, the silences in all—some kind of black hole.

This breakdown of the holding of black in opposition, or in possession, for an all that is equal and made up of whole ones, authorized by an all-encompassing authority, happens through the doubling of vision, the blurriness of seeing, that is announced in glossy black squares.[24] Adu-Sanyah's square is made with a reflective, glossy, and uneven material that does not let your gaze settle. This black (w)hole moves and dances like Taylor across the rectangular piano keys.[25] *White Gaze II Black Square*'s glossy redoubled white gaze, antiblack sight, is what Reinhardt does not want because it is where art, in black, breaks down as self-contained. As Moten elaborates:

> Reinhardt dislikes glossy black because it reflects and because it is "unstable" and "surreal." The reflective quality of the color black—as well as the capacity of the black to reflect—have, of course, been introduced by Taylor. Only now, however, can these issues be addressed by Reinhardt on his own high level. Glossy black disturbs in its reflective quality. "It reflects all the necessarily social activity that's going on in a room." But this is also to say that glossy black's reflection of the irreducibly social is problematic precisely because it disrupts the solipsism of genuine intellectual reflection that painting is supposed to provide. Glossy black denies the individual viewer's absorption into a painting that will have then begun to function also as a mirror, but a mirror that serves to detach the viewer from the social and that characterizes that detachment as the very essence of intellectual and aesthetic experience.[26]

This nothingness as blackness is blackness's excessive aliveness that is breaking out of the square.

Moten's refusal of the mirror reminds us both of Hannah Black's revisitation of Reinhardt's black paintings, as well as Frantz Fanon's problematization of the white gaze.[27] This mirror is revealed by Uzor in *8′46″ George Floyd in Memoriam* by way of the refusal of reflection as a possessive gazing. First in the initial statement that there is the possibility of the mirror as a way

Figure 7.4. Ad Reinhardt's *Abstract Painting No. 5*, 1962.

to claim oneself as both an object and separate, transcending, that object (that is of course the reflected image). In the second statement, the mirror steals away the reflection itself—for the second statement of silence is mirroring the mirror or, we might say, revealing the act of reflection. Similarly, *White Gaze II Black Square* in a first step reflects like a mirror—here only whiteness is seen. This we can understand both as the way in which white light is reflected (due to the process and materials involved in the making of the artwork) but also through the very idea of the black painting, or the black square, as the containment of blackness—as object to gaze on and make a proper subject in contradistinction to. But in her doubling of the reflective act (black square by reflection/mirror), something happens that Reinhardt fears as glossy and Taylor explains as a twoness that "Black" asks us to face.

Taylor's elaborations, which move both alongside black as empty abstraction and black as historically and socially situated, make us hear art as disruptive. Taylor states:

> I think for my first statement I would like to say that the experience is two-fold and later, I think you'll see how the two really merge as one experience.
>
> "Whether its bare pale light, whitened eyes inside a lion's belly, cancelled by justice, my wish to be a hued mystic myopic region if you will, least shadow at our discretion, to disappear, or as sovereign, albeit intuitive, sense my charity, to dip and grind, fair-haired, swathed, edged to the bottom each and every second, minute, month: existence riding a cloud of diminutive will, cautioned to waiting eye in step to wild, unceasing energy, growth equaling spirit, the knowing, of black dignity."
>
> Silence may be infinite or a beginning, an end, white noise, purity, classical ballet; the question of black, its inability to reflect yet to absorb, I think these are some of the complexes that we will have to get into.[28]

Later on, Taylor elaborates in conversation with Reinhardt:

> REINHARDT: Well, I suppose there's a general reaction. I suppose in the visual arts good works usually end up in museums where they can be protected.
>
> TAYLOR: Don't you understand that every culture has its own mores, its way of doing things, and that's why different art forms exist? People paint differently, people sing differently. What else does it express but my way of living—the way I eat, the way I walk, the way I talk, the way I think, what I have access to?
>
> REINHARDT: Cultures in time begin to represent what artists did. It isn't the other way around.
>
> TAYLOR: Don't you understand that what artists do depends on the time they have to do it in, and the time they have to do it in depends upon the amount of economic sustenance which allows them to do it? You have to come down to the reality. Artists just don't work, you know, just like that—the kind of work, the nature of their involvement is not separate from the nature of their existence, and you have to come down to the nature of their existence. For instance, if they decide to go into the realm of fine art, there are certain prerequisites that they must have.[29]

Taylor reveals the twoness of it all. That is to say that the questions Taylor asks are not so as to solidify a cultural sphere among others but rather the questions are disentangleable with the position of Blackness in the world. Out of this then we may derive the fact that there is no such thing as a solid cultural space outside of lives as artistic choices—playing music is painting us. In this sense, Taylor is pointing out not only the antiblackness of a certain conception of the artwork/artworld but also how blackness always bespeaks asking questions that reveal (in their form as questions) the conditional nature of things. In this sense he might be agreeing with Reinhardt that art comes *before* the world, but not as some kind of ex-nihilo creation, where making art is always a claim over nothing and things relegated to nothingness. Rather, he is pointing out how in this absorption of black is revealed a general principle of the disruptive abstraction that art(/black) performs on the world exactly because it *is* the (underground of the or the making of the) world.

What Adu-Sanyah's *White Gaze II Black Square* reveals is how black is always already double, and that the white gaze is the attempt to hold such (falling/failing) off. The double is the movement of both reflection and absorption that blackness allows for. Albeit reflection must be claimed as its inability. For reflection is the scene of policing identification—the claim that that over there is not me, therefore I am, for I can both see me and separate me from this object that proves my existence.[30] For by way of the mirror, or the white gaze onto the black square, whiteness as property is articulated: through holding (in separation, or possession) what it is not.[31] But by doubling down on the black square's blackness—what Moten calls its glossiness, the mirror's doubling, which is the fact that it reflects—Adu-Sanyah is opening us to the fact/experience of blackness. That is, under its glossiness we come to understand how giving up the certainty of knowledge and existence claimed by holding blackness in a square, in an other, is the revealing of the uncertainty that is us.[32] If the glossiness in one view makes blackness disappear, for it reflects white light, then on the flip side it shows how appearance is always already untenable. The appearance of us, as the falling into the artwork that a certain glossiness of it all bespeaks, is the notness of appearances that emerges in their appearingness.

In other words, if it is the absorption of blackness that is needed for the reflection that can be separated, to become an individual that may hold disappearance at a distance from its own existence, then it is the glossiness, as the doubling, that bespeaks how absorption always already took over reflection.[33] This means that the reflection, the identification of an other, of blackness, as that scene where selves are made proper selves, requires the absolute absorption of blackness as the foundation of being able to gaze onto blackness without it being able to do so itself. On the other hand, the absolute absorption of blackness then also bespeaks how reflection breaks in this very claim to be able to control gazing—for it requires an unreflective surface to reflect itself from, where reflection means creating an image that can be held at a distance. The appearance of the self (made object) is arrested so as to claim transcendence over it, which makes the self into something that exceeds the appearance of it (i.e., omnipotent). The claim to hold blackness in the square, the artwork on the wall, the other (as black) outside of selves, absolutely, is the claim to a self.[34] But at the same time, this movement can only be claimed as a narrative/teleology by controlling the way it may be told, which is the controlling of where blackness ends, which is the control over reflection/absorption. What if this scene actually bespeaks the endlessness of absorption that encompasses even reflection—for this is how colors may show. What if it is indeed blackness's absorption that is moving beyond the frame when reflection is attempted to be arrested, and only the reflection's momentary existence finds an end in the square hold of this gazing?

White Gaze II Black Square is part of Adu-Sanyah's series *Inheritance: Poems of Non-Belonging* and is partly composed of autobiographical and abstract artworks.[35] A "German-Ghanian visual artist based in Zurich, Switzerland," Adu-Sanyah's *Inheritance* series elaborates like Uzor's *8′46″ George Floyd in Memoriam* in a first appearance as white and black that is then heard as blackness, as noisy whites and black noise as the black square becomes poetic nonbelonging that is also the recognition that we make belonging together with others, with the world, not as a presence but as a series of holes, of silences, that tie us together to the unknown that is more than all.[36] As Hannah Black notes: "This is not a sociality amongst subjects but some kind of possibility internal to the hole or fissure in the subject, a function of the subject's incompleteness and of the incompleteness that constitutes the subject."[37]

Antiblackness is an imposition that cannot last because it is self-destruction: the separation from the artwork that founds the self is the formation of a death for the self. A break in the white gaze expressed as doubleness, as double consciousness, as not oneness, is the entrance to the celebration of black noise. Adu-Sanyah's square is where art breaks down as self-contained, where lives come forth as Black, and more than one, as less than it, because it is our notness that brings us close to each other and the cosmos, like the blackness (out) of the squares do.

8

Listening with Black Switzerland

Black Box Inferences

We've found something, in our Afrofuturist digs, in our interstitial listening, in our protests, in our counterinsurgent associations. While found, it was already there; while found, it has to be searched for again and again. We might call it an instrument; we might call it a name; we may call it a vehicle, like a mothership, like a black box passed down through time, some kind of ancient code. We can hear now how *black* is some kind of secret ancient technology.[1] Yes, we hear it. Do you hear it? These black marks on pages bring forth an interstitial music, a chorus of voices.

At the beginning of this book, I explained that *Black Swiss* bespeaks something that does not exist—it is/was something we might call *empty and silent*, as in outside of our possibility for thought. Our music as the empty space between words, or proper knowledge, opens our ears slowly, over the course of a musical experience that is this book. If my task is to make Black lives matter, then it is not so as to enter them into an antiblack world of mattering that is built on the owning and killing of Black/blackness. Rather, here I show

how the silence after a word, the empty abstraction behind a name, is before and after the word. Our thought, that is our language as well as all thinkable things, or the structures of our world and us itself, emerges from this emptiness. I will go even further: they are this silence; sounds were always silences and not the opposition to silence. Our screaming protest, our sounds, is the silence of our silenced ancestors, is the unthoughts that we cannot yet speak for, is the unknown onto which we open ourselves.

Black and its other names appear over and over again in the world as such opening of language, as the underground of language that is signification but also grammar. *Black* bespeaks a break in a structuring paradigm. The choice of this word reflects not only people but also the refiguring of what has been claimed to be contained and delimited by making *black* its other. And in fact, the very naming of Black as people emerges as the refusal of a certain conception of lives. In this sense, while the delimitation of blackness is the foundation of an antiblack world, Black bespeaks the refusal of such delimitations of who our people are or may be (and here I do mean this as a question of kinship and belonging). Black, even as an identificatory category, has always troubled an antiblack world exactly because such world must continually control what black may mean, since the continual control of blackness is its (erased) founding principle. John Jacobs Thomas, an author from Trinidad and Tobago, describes in his book *Froudacity* (1890) the way in which in the very moment of slavery, in the allegorical "womb" of the slave ship, Blackness already presented, embodied, and performed a radical refiguring of the grammar of the world by way of the alternate conception of kinship and relations.[2] Black study means opening our ears to unthought sounds, voices, and stories; it means refiguring our listening.

Structures of Listening

If identification not produced and identification not sufficient are used to police Black lives, foreigners, Indigenous, and people of color, it is through another kind of listening, an alternate way of thinking names and identity, one that crosses over the borders for a proper world of authorized citizens who fight and make relations by themselves, that we may listen with and sound for Black Swiss lives. To hear this music and to make these sounds we must thus refigure how to listen so we might find our sound. Listening can follow the methodology of the social register, antiblack identification of threats, and improper blood—or, take another way, one that even refigures what a method can be, one of the city directory, one of the mothership, the listening with the unknown that is our journey through the cosmos on spaceship Earth. What is needed is an acknowledgment of how listening cannot be absolutely present or absent if blackness is involved. We are problematizing listening, retuning our ears, so as to be able to listen without certainty, some kind of uncertainty principle of listening, something like fugitive listening, a listening that hears across borders of what can be.[3]

To begin our elaboration of listening, I move one last time by way of a critique of the proper, of an authorized listening, an absolute registering that delimits itself from the unknown and life. In digital technologies, listening is oriented around absolute measurements, around the ability to know what something sounded, or spoken, means or, at the very least, what consequences shall follow a certain spoken sound, to be able to put listening to work for capital gains. Hence, listening technologies classify sounds, for example, as speech or music, and then, furthermore, must understand such categorical sounds in a relation to other sounds. While we have elaborated on how technologies are used for border policing and how biases against Black lives are reproduced within them, we have not unpacked their methodologies and the structures upon which they rely. To this aim, I unpack the structural biases of listening technologies

To capitalize on speech, speech must be legible not only as speech but as speech that *makes sense*, speech that can then be answered. There are thus two steps involved in a listening that is absolute, that work in tandem with each other: a registering as (proper) speech, and a making sense of such speech (it must be intelligible). It is the legibility that marks speech as proper speech, and it is, in turn, on the basis that speech makes sense, that the machine knows to listen, to understand what is being spoken, which prompts it to answer a demand or relay it to someone/somewhere else and record it into a database.

How does a digital listening technology understand what is spoken as spoken and then respond accordingly? Kate Crawford and Vladan Joler provide us with an exemplary, in-depth elaboration of the networks of resources involved whenever Amazon's Echo, a listening AI technology, engages in listening. The purpose of Echo is to relay the execution of numerous tasks upon a user's request, a request made through speaking. This can range from buying goods to finding a musical work and then playing it back; Echo understands speech and converts it into actions executable by its connected digital software. As Crawford and Joler point out, each instruction or question posed to Echo engages a vast network of resources. Firstly, there is the training that goes into the AI technology, which is composed of all the users of this technology who through their continued interaction train the AI, as well as user data mined elsewhere on the internet used for initial training.[4] There are the raw materials needed to fashion microphones, loudspeakers, and computers, which are mined all over the world.[5] Importantly, many of these resources are mined in Africa, Asia, and South America. And what's more relying on the extraction of resources from the Black world, as most of the cobalt for modern batteries are mined in the Democratic Republic of the Congo. These mining endeavors cost not only in the form of capital but also in the form of lives—lives of miners lost because of unsafe working conditions, children born with health problems because of pollution, and rivers and crops destroyed by mining dust.[6] Resources are collected disproportionally, as a cost to Black lives. But resources also have to get somewhere to make use of them, to then be able to make listening technology. This happens via global supply chains and in the digital realm via data tracking systems.[7] The collection of resources reminds us that data—collected listenings and measurements—also has to be stored somewhere. Servers with all the data found on the internet are deposited somewhere. This collection of resources, leading to an amassing of resources, points to how listening as resource-intensive does not play out on neutral ground: These structures that Crawford and Joler poignantly elaborate, and whose planetary destruction they demonstrate, are centered around antiblack systems of control and extraction. They continue the resource extraction from Black lives and black spaces, started in the colonial period, and are continually rearticulated antiblack systems in modernity.

In this sense, our question of listening for and to Blackness in Switzerland has taken on another dimension: not only is measuring in absolutes unable to speak of Blackness (as and in Swissness) but this kind of listening, or registering, is itself based upon, made possible by, resources extracted from Black lives. Resources—whose extraction destroys our planet, as well as lives on it, disproportionally taking from Black lives—are put to use to listen and understand speech as proper, as authorized, and as controllable. This is about the resources used to control life, which itself destroys lives and identifies what falls outside of proper speech, proper soundings, and poses a danger to the authority over listening and sound.

The relationship between collecting resources and understanding, as well as this being a basis for what constitutes understanding, is what Jacques Derrida identifies as the modality of the archive in general:

> It does not only require that the archive be deposited somewhere, on a stable substrate, and at the disposition of a legitimate hermeneutic au-

thority. The archontic power, which also gathers the functions of unification, of identification, of classification, must be paired with what we will call the power of *consignation*. By consignation, we do not only mean, in the ordinary sense of the word, the act of assigning residence or of entrusting so as to put into reserve (to consign, to deposit), in a place and on a substrate, but here the act of *con*signing through *gathering together signs*. It is not only the traditional *consignatio*, that is, the written proof, but what all *consignatio* begins by presupposing. *Consignation* aims to coordinate a single corpus, in a system or a synchrony in which all the elements articulate the unity of an ideal configuration.[8]

The archive is deposited somewhere, physically (a matter of fact often neglected with digital archives), but is also in the position of hermeneutic authority given to it by itself—it claims self-authorization. Furthermore, the archive's authority requires that it builds *one* "single corpus" or body. But it is not limited to one thing but rather a myriad of parts, or resources, gathered together as, and authorized by, being parts of this overarching *one* body. It is in this latter point, the act of gathering, where something of the hidden process behind a single body is revealed: bodies are made from gathering. Bodies are shaped through the active holding and collecting of parts that form it, which themselves only become parts within this act of collection. This bespeaks immediately both an outside of the body of the archive, where resources are collected, and thus also a specific relation to this outside—one of control. It is a scene of border management that erases the outside by making it mineable. That which falls outside of the listening archive, a body of collected signs outside of interpretation and authority, is absolutely unknown and only emerges when it can be turned into resources, as parts, for bodies, as part of the archive.

Let's think this through with the listening technology Echo. If the act of judgment is one between what makes sense and what does not, as the mining of resources for an authorized inside that rules over interpretation, then the archive (and listening) works like the social register: sovereignty is assigned according to kinship through the distribution of the sign of proper speech, of having a voice, of being a citizen. There is a relation between blood kinship and a sort of kinship of signs, or language, or names, that is based on sovereign authority *named* through being able to own property (a place, real and/or metaphorically speaking, of deposition of resources).

Derrida hints to this: he elaborates on how both the death of the (name of the) father and as conservation of the (name of the) father is of the archive's methodology, as this gathering together under/for one. The father being, here, the wholeness of the archive. As a matter of fact, the archive is both death and resurrection of the name of the father.[9] The archive at once kills the "father," which is the same as killing itself, only to form itself anew. This way it can appear as a band of brothers (those who carry the name of the whole/father), which is a fragmented whole.

In the archive's movement, or narrative form, is also where it becomes clear that the archive requires an outside that it can exercise control over (that something that is outside of it). The archive claims its authority in being able to erase the unknown by taking it into the known whenever it appears—an act of claiming authority in the policing of its limit. Whenever there are holes in its knowledge, whether that be through new knowledge or forgetting, it is able to claim authority over making sense of such outside or over remembering what was forgotten. Thus, the archive at once can be fragmented and whole, acknowledge an outside and remain omnipotent.

Of course, the holes of the archive, its outside, also appear in this narrative of death, which is the distribution of it (and its authority) among brothers. Thus, it is always already exposed and of the unknown. Because its destruction and re-creation must be self-authorized, or self-effected, it must erase the unknown through taking in. Which also means that the passage of destruction is an outside, in a temporal sense, that it has control over. In other words, if we take the space of its destruction as an outside, a space that is not it, then it is clear that if the archive can make itself anew, as it claims, it exercises authority over this moment. This narrative form, that is also a spatial one, bespeaks a structural relation where the inside, or the archive, receives its authority through the control over this other moment/space, there where it is not.

Here the relation to W. E. B. Du Bois's poignant elaboration between laws and chance in "Sociology Hesitant" becomes evident as a question of the inside defining itself in contradistinction to the outside or the unknown (see chapter 3). Du Bois teaches us how the archive, or the space of knowledge, fabricates a space of lawless chaos as its opposition, as its outside. But under the heading of "Sociology Hesitant," Du Bois asks us to think in excess of an authorized and always pregiven archive, or whole body of knowledge, and a separate space of the unknown that cannot be thought. Rather, he proposes to think from the unknown, how the archive is nothing other than what I might call *sounds of the unknown*.

The archive is all of what can be said, but also all that can be heard.[10] As evinced in Amazon's Echo, listening is also entangled in this quest for control over what can be told. But the archive is a narrative structure; the archive is a story, and Du Bois's "Sociology Hesitant" undoes it by shifting from the archive's claim to controlling its borders to an expansion of the underground of the border, that which the archive tries to police with borders.[11] The archive's story, its form, is entangled in (and plays out) sovereignty as it also pertains to citizenship and the nation-state.

The archive and sovereignty create an inside that is proper and authorized to have power and control while creating an absolute outside it has control over, which is also how it claims its authority and properness in the first place. But behind the outside, the unknown that the archive tries to oppose to control, lies something that remains thus always erased. In other words, every time the archive receives new "unknown" information, it claims to control it by turning it into knowledge. Just like Amazon's Echo, which whenever there are new sounds, they are always already placeable inside its all-encompassing structure and set of categories of sound. In other words, all sounds are always heard in relation to prefixed categories of sound. This narrative form, of the unknown being turned into knowledge, is how the archive makes itself inevitable, making its authority omnipotent. It is like the border zone, where all life that comes in must inevitably turn proper and never existed otherwise. This is where "Sociology Hesitant" shifts the narrative. It asks us to think with the unknown instead of erasing the unknown by taking it into a knowledge structure that then remains unchanged. And by revealing the narrative structure of the archive, a moment where the archive ceases to be omnipotent, where it loses control, comes to the fore. This is about a shift in how we listen. And this is as Du Bois announces it, opened by the study of Black lives and thus also of those lives that become relayed to the border zones of the world we live in by this structuring paradigm. It is the study of blackness as that which bespeaks how knowing must change, how our sound is a silence from the past.

The archive polices its borders.[12] Listening is policed, knowledge is policed, data is policed. Think about this triplet of border policing and its relation to identification as a technique of antiblackness, how the identification of Black lives, or blackness in general, is a violent imposition of the law that makes the law omnipotent. It gives the law and the archive, and the state that is sovereignty, its authority. But on the flip side, this also means that those that it polices are unknowns, and, of course, the way that immigrants are called "aliens" in the United States, for example, bespeaks this. It must control those unknowns and turn them into knowns, proper life. But something happens to the archive here; it becomes changed by an acknowledgment of its sourcing of resources, albeit policed. It is the very study, as our narrative incursion, that shifts something in the archive, in what can be heard. By pointing out that Amazon's Echo collects resources, something already happens to listening; we hear it differently. By uncovering this unknown that is required to be mined, something happens to the archive. It seems to become a little more ghostly, a little more strange, a little more spooky, because it seems that its presence becomes blurred by the unknown, by absence.[13]

But at the same time, this can turn also into another erasure of the unknown, another policed border.[14] Just because we know that our knowledge is organized and enmeshed in this vast global network, we don't necessarily know its structuring paradigm, and once we know its structuring paradigm, we can still see it as all there is. This is like the three steps of antiblackness Du Bois already laid out (chapter 3). Let me put it in relation to AI technology: we might uncover the way that there are biases against Blacks in AI technologies. We might then furthermore uncover how the very techniques it uses rely on and are enmeshed in a network that is itself antiblack. We may then also claim that blackness is already inside this structure or, conversely, that we are already outside of it. But lastly, this all still leaves antiblack structures intact because its authority continues to be claimed and reified since it takes its claim from being able to control blackness, the unknown, and Black lives. The unknown is erased by placing it into a border zone that is claimed as absolutely controlled, which also erases blackness from the inside, turning the inside into proper life. In the archive, this proper life would be knowledge itself, and both *proper* and *knowledge*, here, bespeak already an erasure of their own unknown or improperness. But such erasure, as we've learned with the case of Switzerland, can be deferred. Narrative form is how the archive does it. It keeps deferring its unknown into the past as an exception, into other spaces, into an outside separatable from and overcome by it.

Jamaican British sociologist Stuart Hall brings us closer to how we may think about the question of how to tell our stories, the stories of Black lives, in spite of an archive that makes itself in the very erasure of these voices. He brings to the fore that there are voices that make the archive, that the archive's policing happens through those who define it by being able to speak as and for it: "[Michel] Foucault does suggest that an archive is inevitabley [sic] heterogeneous: but it cannot simply be open-ended. It does not consist of simply opening the flood-gates to any kind of production in any context, without any ordering or internal regularity of principle. He does, however, argue that it is not possible wholly to define an archive from within its rules. Partly because—especially in AAVAA's [African and Asian Visual Artists' Archive] case—the very practice of putting the collection together is informed by practitioners who are themselves active participants in defining the archive."[15] Thus, the archive controls its borders. The question arises what "active participants" are; what qualifies them, or allows them to become active? These are the questions that the archive always already answers, or forecloses, to police its borders. In other words, only those who are proper may be active participants. In the same turn, there is thus also always this moment where the archive is revealing how it is not omnipotent, and it is exactly in this moment when it must police that this comes to the fore.

There is a certain similarity in structuring that ties the archive to sovereignty as citizenship. But more than similarity, the authority and authenticity of speech and sounds, or the archive in general, is tied to participants as professionals, as parts of sovereign fields of knowledge. The structuring of oneness as authority, even when such oneness is in excess of one and becomes multiplicity, is predicated on the ability to control its border zone. It is through the erasure of its own foundation, through the transgression of its own fabricated end, that the archive attempts to claim its authority, its sovereignty.

The Cipher of Flesh and Skin

The inside (the archive and sovereign nations) uses the ask for identification as a way to mark the unknown. It does so not because it *creates* the unknown but because it is attempting to separate itself from such unknown, making itself the space of knowledge. In this sense, the inside is unmarked because it is absolutely transparent: it is all knowledge and fully known to itself. The ask for identification is the claim over identification. By demanding identification, this inside is claiming to know what identification is or can be. It presupposes fixed categories; it also means that everyone who is asked is a border zone of its knowledge, and through asking and already implying an answer, it claims control over such borders (or over the outside through making such outside into borders). But when Black lives call Black lives, associate with each other and others in excess of their supposed incapacity to do so, they cut across these borders. Blacks claim the name of this archive, of this sovereign one, of Switzerland, and by spelling it out, *Black Swiss*, they are making an incursion into the authority over identification as registering—it is an interstitial listening. This is a story we can tell through bodies, skins, and flesh, a kind of hoodoo, voodoo, vodun kind of magic, a Black magic, where the masks aren't white, not even the bones, but maybe the clothes, just like the page on which we draw, the sound of drums, "batá," to celebrate orishas in our shango ritual brought on a ship through an ocean from Yoruba.

Let's remind ourselves of the spells we've uncovered in our dig with and for the mothership, this nation under a groove that exceeds the city and flies in outer space where the parliament is funkadelic. They will give us a kind of propellant to fly with the "Mothership Connection." Like George Clinton, we are star children and learned our art of masks from the ancient Blacks, the Dogon and the ancestor gods from Sirius.[16] We uncovered that there is a register of civil society where women's wombs are controlled and their parts are used as measurement of proper belonging. It is through the relegation to the backdrop of femininity, of motherness, that the species is protected from blackness, and such is the ownership of women in the name of the father—of the master of the house, of the sovereign nation. Hortense Spillers points to this by elaborating that gendering happens only within the domestic space, the space of property, which is also the spatial articulation of property.[17] In other words, if gendering happens through the domestic space, then it is also through gendering that a notion of the domestic space appears—that is, women become commodities because their motherness is a property of the house, of the place of collection of resources, and not hers. Furthermore, the very claim to belonging, as the basis for authorized ownership, is instantiated through the owning of slaves, through holding Black lives, through being able to control what falls outside of the proper, through policing the borders of the inside.

Slaves are outside of the body of the archive, of the archive and sovereignty's giving of bodies to proper citizens, of how the owning of a house, a property, as a deposition of resources, mined in the unknown, the owning of blackness, gives bodies to those who own, bodies that can act and have political voices. The slave, on the other hand, could never have a body because the body was already owned by his master, by the master of the house. Spillers elaborates:

> But I would make a distinction in this case between "body" and "flesh" and impose that distinction as the central one between captive and liberated subject-positions. In that sense, before the "body" there is the "flesh," that zero degree of social conceptualization that does not escape concealment under the brush of discourse, or the reflexes of iconography. Even though the European hegemonies stole bodies—some of them female—out of West African communities in concert with the African "middleman," we regard this human and social irreparability

as high crimes against the *flesh*, as the person of African females and African males registered the wounding. If we think of the "flesh" as a primary narrative, then we mean its seared, divided, ripped-apartness, riveted to the ship's hold, fallen, or "escaped" overboard.[18]

The term *flesh* denotes for Spillers the lack of body that is the slave's positioning in the world. Because if to have a body is to be a subject, a citizen, of the state, then the slave cannot have one; they are flesh for proper bodies. But furthermore, the fleshliness of the slave is also bespeaking the production of slaves, how slaves had to be beaten in. Alexander Weheliye clarifies:

> Flesh, while representing both a temporal and conceptual antecedent to the body, is not a biological occurrence seeing that its creation requires an elaborate apparatus consisting of "the calculated work of iron, whips, chains, knives, the canine patrol, the bullet," among many other factors, including courts of law. If the body represents legal personhood qua self-possession, then the flesh designates those dimensions of human life cleaved by the working together of depravation and deprivation. In order for this cruel ruse to succeed, however, subjects must be transformed into flesh before being granted the illusion of possessing a body. What Spillers refers to as the "hieroglyphics of the flesh" created by these instruments is transmitted to the succeeding generations of black subjects who have been "liberated" and granted body in the aftermath of de jure enslavement.[19]

Here, Weheliye also points to the way in which there is a movement of emancipation within the structuring principle of the body in opposition to the flesh. This movement, or narrative, of emancipation from slave to free citizen, from flesh to body, is the narrative of the archive, from unknown to known (and mirrors the narrative structure of predictable to unpredictable and back; see chapter 1). What is thus revealed is that for any*body* there is a need for flesh: bodies must make flesh and claim such as property, or as (which is the same as) overcome, to be bodies.

At this moment, there is another incursion that Weheliye forces us to make, in this account of flesh as *produced*—namely, to ponder how the flesh functions as both before and after of the body, how it moves not only in the space of a *state of exception* (of/to sovereignty) but also before any state—that is, in excess of the temporality announced in the state/body/archive. As such, how can flesh be present without the breakdown of bodies? The naming, or signing, that claims flesh as part, or object as commodity, comes to the fore as this movement of claiming the unknown that the archive continually performs. Here it becomes clear that something is unraveling in Spillers's notion of the flesh—namely, that it is double: flesh is at once made and also unmade. Flesh needs to be before bodies because bodies emancipate out from flesh, but flesh is then also always needed for bodies. To remedy such dependency, flesh is turned into a part. It is the narrative of emancipation that makes flesh into a mere lesser step to bodies. When Spillers brings forth the term "hieroglyphics of the flesh," she reveals this doubling: the way in which the flesh is both there before and in excess of the body of men and also covered and claimed by this sovereign body. The hieroglyphics of the flesh refers to the image of the lacerated backs of slaves where flesh exceeds skin as unmarked proper body. But it is also not reducible to this scene of violence, for the hieroglyphics bespeak an excess of these kinds of brutal mechanisms toward gaining control over the slave. One way to think about it is that these lashes also were used to punish fugitive slaves, those who attempted to (or were suspect to) flee. Thus, these lashes bespeak not only a violent attempt to regain control but also the very resistance to the system of slavery. *The hieroglyphics of the flesh are symbols of the very end of an antiblack world, for in the fugitive comes to the fore the need to protect this antiblack world from its end—the need to control the borders of this world.*

We must conclude that *making* bodies is a process that expunges the *hieroglyph* from the flesh, from sight—a claim to the proper order between flesh and body. In other words, flesh becomes skinned in, and as, the process of covering over hieroglyphs, turning flesh into fleshly part, as commodity that is nonetheless emancipatable and countable. And this is a question of identification as a question of controlling Black lives and Black skin.[20] But it is also where Black lives, and Black skin, escape the negative definition, the being held at the border, of the proper. The antiblack system erases Black skin by making it into flesh but must also erase the fleshliness of skin, the hieroglyphicness of it—the lashes on the backs of slaves—in a second step, because this flesh must be controlled to become parts of bodies. In other words, flesh must be identified so that skin, as that which ties flesh together into bodies, can be erased to form bodies that are whole. Spillers hints to this with a sentence: "These undecipherable markings on the captive body render a kind of hieroglyphics of the flesh whose severe disjunctures come to be hidden to the cultural seeing by skin."[21] There is a relation between the erasure of flesh and a negative identification of Black skin that is also the erasure of racialization.

This equation also requires flesh to be in excess of its erasure and producedness so as to be available beyond the archive/body. What hieroglyphs of the flesh reveal and hide in their articulation of skin when they are flesh as parts is the breakdown of proper bodily composition. Hieroglyphics of flesh speak of a notion of flesh that invades the skin as borders for the making of proper bodies, where the skin is hieroglyphic and hence over-touched and outside of touch, both at once.

Claiming ownership over flesh is the commodification of flesh into parts of bodies, of society, and it is in lashes, that are the making and erasing of skin, that such ownership is attempted to be made (real). But it is in those very lashes (in the need for lashing enslaved Black lives)—which is also a metaphor for all the techniques that an antiblack world engages—that become hieroglyphics of the flesh, that the archiving of parts breaks down and needs to be reclaimed by those who own bodies. In freedom, thus, Black skin is erased; the hieroglyphics of the flesh of Black lives is erased. But hieroglyphs are here more than antiblack systems of control; they are lives in excess of it, they are ancestors, the conditions of the world, and how what is came to be. The hieroglyphs are there before the lashes, before the flesh as part; it bespeaks flesh as more than and beyond parts for bodies. It bespeaks lives as more than and beyond life only valued because it can own flesh, property, and commodities. Remember, *the lashes are there to erase the very hieroglyphs (as the name of the encounter with the outside/unknown) it claims to own through making them*. That is to say that the very concept of the lashes, which includes the instruments of lashing, including but not limited to the act of lashing as well as antiblack structures, aim to erase only themselves—the very need to lash, which is where flesh is claimed as made and separate from the body. What if the lashes continually fail? What if the need to refine the lashing technique over and over, installing new measures of border control, identification of nonproper lives and noncitizens, keeps improving? Because no matter how many lashes, the hieroglyphs keep encroaching in on it. This is a shift in how we tell the story; it is an incursion into narrative, into the form and structuring paradigm of these instruments.

The hieroglyphic reveals something about the process of the sign itself: it reveals that meaning—that is, signs and names—makes itself in relation, and that such relation is absolutely exposed to the unknown.[22] Relations' ruse of authority, given by being of the archive as singular authority under the name of one and as one whole body, falls apart in the very space where it also claims its authority by covering such unknown limit over—the hieroglyphics of the flesh that bespeaks a fugitive sociality that crosses over borders of relation.

Listening, it becomes evident, is not a matter of sound vis-à-vis other sensory stimuli, nor is it about listening modalities that can be chosen at will, but it is rather about this question of policing and identification. But listening also bespeaks the hieroglyphics of the flesh, the way in which this archive is making itself in the unknown. Thus, listening is at once engaged in racialization, control of proper belonging, while also covering over an opening that exceeds all that can be heard.[23] But we must go beyond a notion of it simply being a choice of what can be listened to because such would reduce listening again to an individual with authority over its ears. Because listening becomes claimed as a site to instantiate authority, it is entangled in the very delimitation of life and what can be, what can exist as a sound itself, what can be a voice or a citizen.[24] Listening is also where ownership is articulated.[25] Listening's authority is not simply a matter of how much power it has; it is rather of a structural kind.[26] The archive, as a deposition of listening and speaking, as well as (musical) sounding, is always organized around claiming authority over what can be said or heard. But, as we've uncovered, listening as a border zone can also become this site where the unknown, the underground in excess of any border, can become audible.

Being able to claim knowledge as absolute means holding authority over its borders. It is in this authority over itself that it requires a gap to justify itself, exactly this border, because if speech was listening and flesh was a body then the body's sovereign authority would cease to be exercisable: if the body is omnipotent, then it cannot exercise authority over itself and hence cannot have full authority.[27] One can thus say that there needs to be multiple bodies to justify authority of *bodies*, but this again opens an absence, a space of nobody that roots bodies' authority in exchanges with each other. This, in turn, could then nullify bodies' authority over no-bodies because it would mean that bodies are simply no-bodies. Hence, the body's formation in flesh (no-bodies) needs to be covered. This cover happens through the owning and controlling of borders, and thus also through controlling what can be said and heard, who can have a voice.

Gayatri Spivak in "Can the Subaltern Speak?" elaborates alongside speech's involvement in the political and the impossibility that such poses for those who have no voice in the current political order. As Spivak elaborates in this seminal work, speech is not readily accessible to the subaltern (i.e., those who are structurally excluded) because speech moves in the ideological confines of the world. By this, she means that the subaltern's speech is always accounted for by someone else. The subaltern can't speak for themselves because speech is always already formed within a political system that does not include the subaltern as agents.[28] If the subaltern is not a subject within the archive, then it cannot speak for itself. Rather, it's always already spoken for by those who have bodies, who have agency in politics. That, of course, is not only a function of someone else physically speaking for the subaltern but also of the subaltern's speech (its sense—that is, its relational valuation—as well as its *name*) not being able to express *for* the subaltern because to be a speaker has to have authority within the archive.[29] In her brilliant study, Spivak poignantly elaborates alongside the question of two kinds of representations: one is *Darstellung*, something like presentation and rooted in visual appearance, and the other is *Vertretung*, something more akin to standing (in) for someone in a political sense. Any subject needs to be at once representable within politics as well as be able to be the appearance of politics itself. In this sense, the registering of citizens is not only the recognition of who has a standing in politics but also must present citizenship itself as them.

Within the scene of emancipation in the United States, as an example, many have voiced that the Negro should receive both a political and juridicial voice. Voices are parts of sovereign authority and as such hold sovereign authority, but in the case of the slave, the masters owned their voices, giving them authority—the more slaves someone owned the more power their voice had. W. E. B. Du Bois explains: "But there was another motive which more and more strongly as time went on compelled the planter to cling to slavery. His political power was based on slavery. With four million slaves he could balance the votes of 2,400,000 Northern voters, while in the inconceivable event of their becoming free, their votes would outnumber those of his Northern opponents, which was precisely what happened in 1868."[30]

That this absence of having a voice is also a question of study, and the archive as a question of knowledge becomes evident in Aldon D. Morris's analysis:

> [James] McKee argued that the sociological perspective on race failed because its theorizing and assumptions led to the manufacturing of a Black people who did not exist. American race relations could not be examined accurately when one of the major groups in the equation was so caricatured that its members imitated myth rather than reality. This perspective was not capable of anticipating the civil rights movement or understanding real racial dynamics. The source of that failure, according to McKee, was rooted in the racial assumptions of white sociologists, which they shared with the majority of white Americans. The guiding assumption—belief in Blacks' inferiority—was shared by white sociologists of race during the closing years of the nineteenth century and throughout three-quarters of the twentieth century. It

was responsible for producing a sociology of race that was marred from the start.³¹

Antiblackness demarcates Blackness as an absolutely unknown space so as to claim absolute authority over voices, which is to have authority over what counts, over knowledge and all power. Thus, not only is Blackness marked as outside through not being able to speak, or to be told about outside of a narrative of emancipation that removes the blackness of Black subjects, but Blacks have no voices.³² But as we've uncovered, this absence of having a voice, as an absence of sound, is a kind of silence that cannot be erased because it is the silence that sounds something. It is the very absence of voices that bespeaks a radical danger to a world of proper sounds. For these silent voices are where voices are claimed in the first place. It is this black noise, this silence, that bespeaks an excess of listening, a kind of listening that comes before and exceeds sound, because it is the improvisation that composes other sounds, sounds unheard.

What happens to the archive when blackness voices itself, when silence makes sounds? Can listening become a black art, be study of the black ark, of a black ship that fugitively steals away in its unthinkable sounds as music? If there is in stolen life a certain stealing that flows away from contractual relations, then what does it sound like in its improvisation of screams and silences? In *Betty's Case*, there is a certain stillness that her nonperformance plays in and out from. Her stillness is a silent protest, is a claiming of the right to stillness of birth—I mean that as a decriminalization of stillbirths, which becomes criminalized in antiabortion laws, in antiblack state-enforced brutalities like in Regina McKnight's case, who had to spend twenty years in jail for a stillbirth.³³ I mean that stillness to mean also a loudness, louder than equalizing and equalizers can measure or control, like the screams for George Floyd that are silence's overflowing of itself as (black) noise.

Speaking *Usländerdütsch*

What has come to us in the scene of silence that exceeds its definition (as the negative of white noise) is a listening that comes as the silences in sound, that turns sound into silences, where the silence of listening becomes the necessity for composing sounds. This is where the listener, the observer, falls over into the music, into the artwork. This is like this moment when the police ask for identification, as the nation asking for identification (of Black lives) is countered by stealing their silence from them. If the very ask for identification is the erasure of silence, as the absolute knowledge of the inside and the absence of color from it, then this happens through killing, as identifying, the unknown. It is the claim to owning silence, owning listening and being able to silence Black voices, that antiblackness practices. And this silence is expressed as a lack of listening, as a gap in the ability to hear. Silence is the opening of listening because listening is silence, the silence in sound, and thus it is here, in this space, when the closing of that space is refused by Black lives not showing identification and associating in fugitive socialities across borders that they are Black lives, that there is antiblackness, that unsettles this antiblack world.

Nahum Dimitri Chandler elaborates in his book *X* how an attention to the flesh in excess of its relation to the body brings forth this hieroglyphic cipher that is a fugitive listening toward another kind of cosmic music. Chandler uses passages from Olaudah Equiano's autobiography, *The Interesting Narrative of the Life of Olaudah Equiano*, to elaborate and open this kind of interstitial space. Equiano, after gaining his freedom, became an abolitionist and influential writer.

> The key phrase implies, all the more tellingly because of its ellipsis in style, the risk of absolute loss for [Robert] King that is posed by Equiano's capacity for will and perhaps intention. The paradox is that Equiano is more likely to leave if *not* manumitted. So Equiano is manumitted by King, in order to maintain (the latter hopes) some measure of control over the former slave's intentions, as wage laborer.
>
> This force that I have just outlined is the "double" of the will of the master. It not only delimits his will but specifies it as his, as that of a slave owner. It also organizes and directs it, in the quite specific sense that it forces him to act in the interest of another in order to act in his *own* interest, although we have just complicated any notion of exactly what the form of intimacy suggested by this word of self-possession could mean here. Again, we can come to a distinct recognition of the construction of a Europeanist subject by following the particular and quite specific story of the making of an African American or diasporic subject.
>
> The play that is set afoot is the movement of something that we can position under the heading of "double consciousness" as Du Bois formulated it—seeing "oneself through the eyes of others"—although we might be obliged to say at this juncture that the movement in this case occurs as a kind of double unconsciousness. Not privy to the incipit

of its own position in relation, it would too yet disavow the *force* of the other—the affirmative and active bearing of its practice—within that relation. And the matter would take on an even darker hue—a form of redoubled unconsciousness—as one can theoretically name that the positive existence of the discourse of the free individual in the rise of the modern text of the Atlantic basin is itself a putatively coherent proposition only if such is resolutely presupposed as secondary to an originary and always previous singular whole.[34]

What Chandler is uncovering here is the way in which there is a moment in the narrative form that turns flesh into body, slave into subject, that requires a gap that can be heard as a moment of a kind of erased agency in the slave, a moment that is under the border zone. This moment, a kind of singularity with a wormhole, reveals the beforeness of silence, of flesh, to the wholeness in opposition to the border. It is like this scene of resource extraction, an interstitial space that must exist to claim authority but that also undermines authority itself because it bespeaks how authority must claim itself. How the meeting of the unknown that the archive requires also opens it to the possibility that it might not have control over its borders.

Hall's elaboration of a living archive focuses our attention to this same moment of unpredictability:

> It is impossible to describe an archive in its totality. The very idea of a "living archive" contradicts this fantasy of completeness. As work is produced, one is, as it were, contributing to and extending the limits of that to which one is contributing. It cannot be complete because our present practice immediately adds to it, and our new interpretations inflect it differently. An archive may be largely about "the past" but it is always "re-read" in the light of the present and the future: and in that reprise, as Walter Benjamin reminds us, it always flashes up before us as a moment of danger.[35]

While Hall is reminding us that the archive is colonial, expanding and trying to claim new spaces, he is shifting us to how this also opens it up to a danger, which comes before its control. This danger is the very loss of its authority at the moment when it is re-read, when it must change, when it becomes discredited because it is not sufficient in itself, as it was never in itself. This moment is not an instance in time. Rather, we must hear this structurally. The reduction to an instance makes this moment into an outside—a border that can be controlled. It is the way in which this moment is no moment, this place is no place, how nothing changed, that this kind of *changing* becomes the radical shift in and as this story.

If listening is invested in resource extraction and allocation, in claiming, through judgment, an authority for itself and its parts, then to point to this claiming (that is, also to listen to listening itself) is to move through not only its beginning but also its end. It is something akin to Du Bois's metaphor of the city directory, of a mode of listening to Blackness (in Switzerland) that has to give up knowing as self-contained—that is, knowledge as authority of the archive/listening or measuring. Rather, knowledge of the world, of lives, becomes nothing other than our continual engagement with (and as) the unknown. This is not so as to declare Blackness and its sounds unknowable within Switzerland but as any knowing of/as space and its insides/outsides as never complete, as always in improvisational (dis)forming. Of course, *improvisation* here, in my use of the term, is process and nonprocess, as it has to not articulate something foreseeable nor an unstableness, but rather it is sitting with listening as an uncovering of how what is is always already about what it is not, that which is outside or unknown.[36] Think of, for example, the *foreigner*'s language in Switzerland—by that I mean that which is called foreigner-German (*Usländerdütsch*) in German-speaking Switzerland. Proper speech as national belonging and improper speech as both the sphere of those who don't really belong here and Blackness is a matter expressed in sounds and their measurability in tongues.[37] *Usländerdütsch* is improper speech, articulated in Swiss dialects as well as in standardized German: it problematizes both the proper German tongue and the not-so-proper but still proper dialects of Switzerland as measures of belonging. Foreigner-German is never finished, never definable, and is ultimately language as an expression of lives lived in and across space and place—it has sounds and patterns from other (languages and geographic) spaces. This is something like a Glissantean creolité that describes this unforeseeableness of meetings. We gain from Édouard Glissant an understanding of thinking the unforeseeableness of meetings that is expressed as creolization in speech and lives.[38] Speech without bounds is this improvisational generativity of lives in complex relations, across borders of nations and delimitations of life.[39]

The demarcation of languages, and their use as national and racial markers, becomes unraveled in a Derridean archival uncovering. What happens is that instead of following the authorized presence of the archive and its sounds, we follow the ghostliness of a flesh that exceeds any archive. This is the cipher of the hieroglyphic flesh, the space where we improvise and compose our sounds out from. Behind the political sovereign authority—that is, the shared voice among citizens—is the exception to the one. In other words, the fact that the archive is both one and many, as a body made up of bodies, bespeaks the structuring of the political voice: to be able to speak one must be a voice for and of this one body. There cannot be any voice that does not speak from what is sayable and thus tied together as one voice. But that which this whole takes as its contradistinction, marks as its exception, is the stolen voices of slaves, the silence of the unknown. This is what lies behind the border zone, behind the skin, behind the name of the commodity or its value: the hieroglyphics of the flesh, the no-body, which is also no voice or, we might say, it is no one. This no one is not really the opposite of one nor a proliferation of one, is not no voice nor a proliferation of voices. Rather, it bespeaks how the very wholeness of the authorized oneness of the archive and its parts, of the nation and its citizens, is made in opposition to its own notness, its own end. This means we might now read singularity as a hole, a black hole, which is to hear the interstitial behind singularities, as a breakdown of the one, the count, as counting and as an account, itself that is both an excess and a lack. The interstitial is thus not-in-between—it is the emptiness (or fleshliness) of the whole and its parts.

This is a question of the underground of identification. Derrida speaks about dating, but we might as well read such as identification: "Of a date itself nothing remains, nothing of what it dates, nothing of what is dated by it. No one remains—a priori. This 'nothing' or 'no one' does not befall the date after the fact, like a loss—of something or someone—nor is it an abstract negativity that could be calculated here, avoided there."[40] Derrida *dates* the name and names the date of its ending. It is the need for the deferral of this nothing, of this not me, that which is not the identified, that turns into antiblackness. Antiblackness cannot last because identification always turns into not me, into me not being me by myself, this nothingness of me, that bespeaks me as always with others, me as always of not me, me as the expression of the ancestors, of the notness that we hear as our past. In other words, antiblackness's need for continual identification, which both delimits the unidentifiable as black and at the same time identifies it, thus erasing this unknown, is itself untenable. For this need to ask for identification—that is, the circumscription and containment, and the performance of, the unidentifiable—is existence's own end. And this end is already there before it and has never left. Thus, the unknown needs to be deferred onto others—that is, needs to be localizeable somewhere else or, more precisely must remain as absence. Furthermore, this absence needs to remain absolutely absent—that is, separate from what is here, what is already claimed as part of the sovereign domain. This requires the fashioning of a border zone, which is the erasure of the border, or end, of this domain through exercising control or authority. But this also means that the unknown, or blackness, is inevitable. What is thus antiblackness other than a narrative trajectory, a kind of authority claimed in the control of the direction of energy, or resources, that is symbolic and material—which nonetheless makes it exist, for narrative here is not limited to texts but rather life and the human world's structuring of it. This border is thus also the opening, the revealing of the way in which the grammar always fails because of the fugitivity of the emptiness that is behind abstraction.

This movement of nothing is the underground of a chain of authorized relations, the mothership—that is what "blackness/Black" appears as.[41] What lies beyond the chains, like between the shackles, is simply the proliferation of nothing as us, the no-ones that is lives living together with those who are not here or us. Listening becomes a continual getting-in-touch with the unknown, with the nothingness that is our togetherness in traveling through space-time.[42] We could call it an uncertainty principle of the ear, of listening, and as such, it is Du Bois's hesitancy and his sociology.[43] It is listening in the suspension of judgment, judgment as delimitation of the sensible and sense through antiblackness and its techniques. Blackness's overflow from judgment is improvisationally listening where listening cannot *take* place, own land or people, because there is no-one to listen to. Listening becomes our love for what we can't hear.

Derrida can provide us one last text to mess with in our unlistening agenda of black noise/silent speech/silent listenings. What he points us to, after our elaborations, is how the underground that blackness bespeaks continually exceeds the recognition of it. In other words, blackness indeed requires constant refiguring, for it is the unname, as the process of continually *finding* new thought, new words, new texts, and the continual work of exceeding what can be held/known. He writes:

> That justice exceeds law and calculation, that the unpresentable exceeds the determinable cannot and should not serve as an alibi for staying out of juridico-political battles, within an institution or a state or between one institution or state and others. Left to itself, the incalculable and giving (*donatrice*) idea of justice is always very close to the bad, even to the worst for it can always be reappropriated by the most perverse calculation. It's always possible. And so incalculable justice requires us to calculate. . . .

> To keep this from being a truism or a triviality, we must recognize in it the following consequence: each advance in politicization obliges one to reconsider, and so to reinterpret the very foundations of law such as they had previously been calculated or delimited. This was true for example in the Declaration of the Rights of Man, in the abolition of slavery, in all the emancipatory battles that remain and will have to remain in progress, everywhere in the world, for men and for women.[44]

What if Derrida's mentioning of the Declaration of the Rights of Man and of the Citizen, and the abolition of slavery as exemplary moments, reveals something in the underground of these scenes? In other words, if these scenes become legible in a narrative of progress or emancipation, then they fail to bespeak that which these terms denote (or maybe we could say these names, as an entrance into a grammar, are always already such a coverup, albeit this critique becomes untenable after our elaboration). What if it is a question of how a grammar is also an *apparatus* of narrativization?[45] That is, if the incalculable is blackness, and if the control over the incalculable is antiblackness's absolute authority, then the uncertainty principle exercised by opening the question of what us and the world is, is the opening of grammar as the refiguring of the (non)relation between what is and what is outside of relation. It is to reveal how the grammar that proclaims control over rupture has always been a lie. That is also to say, this kind of scene of unanswerable questions is opened by Black lives protesting loud as well as by their silent refusals to the demand of proper identification. The uncertainty of knowing, the openness to asking questions again and again, is the opening of the world to the impossible that we call the unknown. To listen again and again, from the position of the uncertainty of hearing, the not knowing if there is sound, is making music, is listening for unthought voices—unimaginable musics. Beyond the antiblack gaze that is holding our ears, so as to establish who belongs, as a controlling of what may sound, of what is us, lies a bizarre out-of-placeness that a doubling of gaze, of listening, announces as an out-of-space flight. Fly through a door of no return, to where Blackness and lives steal away as lives lived in a musical black noise ensemble.

9

Interstitial Listenings V
Charles Uzor's *Bodycam Exhibit 3*, Part II

Enter the Black Hole

We ended our first engagement of Charles Uzor's *Bodycam Exhibit 3: George Floyd in Memoriam* in chapter 1 with an opening in the middle of the narrative form: equilibrium to disequilibrium and back, where the middle bespeaks the loss of control over Black lives. And this moment of disequilibrium is also this moment of refusing identification, this showing of a mirror to an order that tries to control Blackness through owning borders and delimiting what life inside may be (see chapter 2). This loss is a kind of black hole; it is like this singularity where the laws become both revealed and break down, and maybe there is a wormhole, a kind of hidden passageway to worlds unknown. In some ways this book became this passage, this flight with the mothership through the hole in all there is, to hearing the cosmos anew. And it is here, where these interstitial listenings exceed their containment, that the book comes to an end with our falling into this empty nonspace.

What did we learn from our dive into interstitial listening, our study of blackness? And as a result, what might we hear now in *Bodycam Exhibit 3* that we haven't heard before? In what ways has our study brought us new ears, another way of listening, one that continually opens our ears, again and again? Blackness cannot enter the music, the world, that is defining itself in antiblackness; that is, a series of aesthetic choices as abstractions absolutely separatable from life, nor a reality of proper life. Blackness cannot enter music that sees itself as complete by itself, as untouched by social lives, as purely abstract, and, in the same turn, Blackness is only an abstraction for a world where what is real is only what is legible. Blackness is the backdrop against which both a good abstraction and a real proper world is distinguished. But listening from the mothership that flies in wormholes, the holes of it all, we hear another story, one of an association between abstraction and realness, a fugitive movement that we might call the continual refiguring of relations with the world.

Bodycam Exhibit 3 problematizes a notion of pitch and an extension of music, as absolute abstraction and a separate real world that is completely outside of music. It does so through mirroring, blurring, and shifting, where listeners are confronted with their own listening—it's like this protest of refusing identification. Two things happen to pitch with Uzor's compositional technique in *Bodycam Exhibit 3*. One, pitch's neutrality is problematized, and two, pitch becomes a method to practice what might be called a *wake*—a practice of remembering those Black lives lost to antiblack brutalities while, in the same turn, celebrating lives lived in and with Black.[1] Pitch itself, as sign or as names, is problematized in its sounding and writing as a carrier for abstract musical objects. Recording or documenting reality happens through musical pitches. Of course, this recording is not just a capturing of a pre-given real but is entangled in the forming of reality through it being always also archival material—the archival material is involved in shaping how we listen to and think of the world and, in turn, materially affects what happens to and in the world. If the recording as a listening technology has to be scrutinized regarding its conditions of existence because of how it reproduces antiblackness, then this is not just in the form of a question of proper capturing, which as a discourse remains within a quest for objective and neutral fidelity. Rather, the forms of dissemination and how recordings are *consumed* have to become a part of the question of recording. Technologies of recording are listened to by someone, and who does the listening influences *what* is recorded and *how*.[2]

This is evident if we think of the biases in photography—the biases appear because of an assumption that the humans imaged will be white (and not Black). Simone Browne's seminal study on surveillance technologies in

Dark Matters, from slave branding to modern biometrics, reveals a deep entanglement between race and surveillance technologies, not only photography but also its digitization. In her discussion of iris scan technology she points out:

> Prototypical whiteness in biometrics is an extension of the "general culture of light" that Richard Dyer lays out regarding photography, film, and art. This is a culture in which, as Dyer asserts, "white people are central to it to the extent that they come to seem to have a special relationship to light." The logic of prototypical whiteness is seemingly present in earlier models of iris-scanning technology that were based on 8-bit grayscale image capture, allowing for 256 shades of gray but leaving very dark irises "clustered at one end of the spectrum." The distribution of this spectrum's 256 shades of gray is made possible only through the unambiguous black-white binary; the contrapuntal extremes that anchor the spectrum, leaving the unmeasurable dark matter clustered at one end.[3]

Browne demonstrates a sublimation of earlier technological conditions in photography to the digital realm—technologies, which themselves have been formed within the scene of categorical distinctions of humans/life. What is poignant, too, is that it is the absence or lack of thought given to Blackness with an a priori assumption of the human as white that informs the lack within imaging technology. As Sarah Lewis elaborates, historically "light skin became the chemical baseline for film technology."[4]

This basis for modern border control is thus engaged here in this musical scene through the question of sound, the question of what accurate representation, or capturing, means. In the same manner, as there are biases in visual technologies, sound recording technologies capture with an assumptive logic of who is both doing the recording and the listening of the recording, as well as to what aim. In *Bodycam Exhibit 3*, all the musical pitches are references to the real event, which becomes clearer in their pairing with other imitations such as imitating speech sounds and the positioning of musicians in space. None of these pitches exist without the referent, but by listening to the music, the listener continually reconnects pitches to imitation material and, as an extension, to the real event—because the same pitches can be found in the instruments' direct imitation of the sound recording. That is to say that if the tuba imitates George Floyd saying "George" at the beginning of the piece, then there are pitches that can be identified in this sound. Uzor found through analysis a central pitch and gives it to other instruments to play later in the piece. Thus, our ears will hear relations between these two sounds.

It seems that *Bodycam Exhibit 3* refuses the very possibility of a clear-cut distinction between real and art or music. In some ways, this is announced by the very medium of phonorealism. Let me turn again to Douglas G. Barrett's analysis of Peter Ablinger's phonorealism. The border zone that is continually crossed in Ablinger separates two spheres: one of sound as abstraction and one of speech as the real thing, as documentation of something legible in the archive. What emerges in their case is that music and sound belong to abstraction; sound is not about communication, nor about its physical properties. Rather, it is something that "invades us."[5] It seems at first strange how abstraction is both outside of everything physical, communicational, and is illegible while also a kind of force that "impels" and "invades" listeners.[6] Somehow this is very close to the notion that abstraction is, on the one hand, empty and thus merely a carrier of exchange value, marked by a sign, and, on the other hand, is something both dangerous but also a resource for the value of this separate space that is real and can communicate something. But it is very clear now that this is a matter of border policing: it is the need to make abstraction into an exception, something controllable, that is absolutely outside of the real world, of communication, that relegates abstraction to somewhere else and makes its movement over the border, into the inside, an "invasion." But, in the same turn, this phonorealism work at the border zone that bespeaks how there is this underground of it, in excess of the border as mere control of delimitation, a movement that exceeds the separation between abstraction and real, between not us and us, between music and communication.

The distinction between something that is outside of reality and something that *is* real, as a distinction between something that is unknown and something that is known, is a distinction that relies on a border zone made as that space where there might be an invasion, where abstraction can impel into reality without its consent, where reality becomes abstraction. This border zone is either a border or an interstitial listening, either a control mechanism that erases one from the other or the very excess of relations, where the outside is always in touch in excess of what the inside is. This contradistinction of domains, in the case of the sound recording, relies on a notion of fidelity of documentation entangled in the authority of the archive, claimed in the border zone through being able to decide what is real and what is not, by holding off abstraction and delimiting the real. Music can only be outside of the realm of reality, outside of context—that is, as either overly physical or a lack of matter—through the claim to authority over interpretation and naming. This is the archive's claim to authority by delimiting spaces through borders into something it can exercise its control over. *Listening is a border; it is the border between what counts and what doesn't, what is abstraction and what is real.* But listening is also interstitial, where the border turns into the crisscross that is movements across, queering over lines, hearing anew, tuning in, into sounds unknown.[7] Listeners of this piece are confronted with not only questions of what observation means in this context but also where the line between observer and observed becomes articulated or blurred. The observers, or listeners, and their involvement in this scene becomes unavoidable: either there is a continuation of the separation from it—marking music as separate and the listener as separate, and thus allowing the (re)performance of antiblackness—or the illusion of absolute separation is broken and the listener becomes part of the music, a music that protests because music is entangled with the world of its making, and musical acts matter—not as a kind of collapse between the two but rather as the end of one's authority over the other, over the outside.

This is evinced by the doubling of the police officers, which reveals an erasure of doubling that is also the very prerequisite for the archive's white gaze, its claim to authority against blackness claiming control at its borders. There is a stipulation in Uzor's score that states that whenever someone plays the same pitch, then the musicians must move their pitch up or down, away from the heard pitch. Because the police officers are doubled—not only as characters but also in their musical material—they have more of a chance to end up playing the same pitch, and they must continually avoid this doubling of pitch. Furthermore, the doubles have to look for each other by walking around on stage. Thus, they at once must avoid their double, in sound, and at the same time continually look for their double. To me, this represents the way in which there are proper lives who must be like each other, they must share the same trait, but can never be/meet each other. They *make* themselves like each other through identifying those who are not like them because they cannot do it through identifying each other (through playing the same pitch). This is how antiblackness works. It is the continual deferment of the pitches meeting themselves, their mirror image, that becomes the exchange of them among those with equal standing that also makes each pitch a sign for mere abstraction, something empty to attach a sign onto. Abstraction is held by the signs that marks an ownership of abstractions and gives those who own these signs standing in exchange of relations. Thus, the attribute that makes equals is erased and at the same time essentialized—which is the same as erasing essence or individual instance and/or conflating essence and instance. In short, it is the separation of essence and instance, turning every instance into essence. This is the normalization of society that professes absolute diversity and is built on the exclusion of those who are not of this essential trait, those outside.

We might pose this as a question of musical form, as a question of the relation between instances, or content, and form, and/or essence. As Theodor Adorno states:

> "Sounding" and "in motion" are almost the same thing in music, and the concept of "form" does not explain anything about what is concealed, but merely thrusts aside the question of what is represented in the sounding, moving context that is more than mere form. Form is only the form of something that has been formed. The specific necessity, the immanent logic of that act eludes the grasp: it becomes mere play, in which literally everything could be otherwise. But in truth, the musical content is the wealth of all those things underlying the musical grammar and syntax. Every musical phenomenon points beyond itself, on the strength of what it recalls, from what it distinguishes itself, by what means it awakens expectation. The essence of such transcendence of the individual musical event is the "content": what happens in music. If musical structure or form, then, are to be considered more than didactic schemata, they do not enclose the content in an external way, but are its very destiny, as that of something spiritual. Music may be said to make sense the more perfectly it determines its destiny in this way—not only when its individual elements express something symbolically. Its similarity to language is fulfilled as it distances itself from language.[8]

Adorno is pointing us to thinking a more complex relation between abstraction and representation, where music is a kind of border zone at a distance to language. But what if we do not follow this syntax, the kind of overdetermined destiny of content, as it unfolds as form and more than language? By that I don't mean the forgetting of the structuring grammar but rather listening to the underground of syntax, the kind of excessive possibilities announced not in absolute chance but as the very excess of every kind of grammar. To put it simply, behind a story with an inevitable end is the writing of the story, the telling of it, that can change its end(s). What if we think in the underground of context, in the backdrop of this grammar, these signs in relation that evade the emptiness, the abstraction?

Space Music

Music is not self-contained but, rather, if it is a border, and if we turn the border over into interstitial space, it exists as that which exceeds any hold—before any hold as well as during and after it. Music forms spaces: the snare drum resonates acoustic space, sounding empty/silent but at the same time, with a shift in listening, when the sound decays, it makes space (audible). If the snare drum is antiblack paradigms of sovereignty in a first turn, then it is revealed, after time has officially ended (i.e., when the sounds or the music comes to an end), that in its unnamedness the snare drum speaks of something radically outside those paradigms. Its sound is always in excess: too loud, too much blurring that cannot be accounted for, too much reverb that shifts timbre, too big for the space of the music, too unmusical to be aesthetically pleasing. Hence, in another direction than linear time, it speaks of a musical space formation that is of an alternate form, another kind of narrative: the decay of its sound not as decay but as spacing announces its articulations, not as cause for space but as space's ripple effect. The snare drum, aptly named "the Unnamed" by the composer, becomes blackness, not as undifferentiated abject other but as *spacing* itself. Thus, in the unnamed that are our silence, that are Black ancestors—that is, the ghostly presence of other worlds and other beginnings and ends—lies hidden lives as more than individual pitches in space.

Music becomes space, and space becomes a meeting place for lives. Remember the tuba's skip in the middle, that moment when the tuba steals away in the uncertainty of the border zone, before it is turned into a border? Let me clarify this moment of leaping grounds. To do so, let's remind us of Nahum Dimitri Chandler's elaboration of the founding of the sovereign subject alongside Olaudah Equiano's manumission: the moment where freedom is given and control is returned to the master is also the moment of unpredictability in regard to what Equiano will do—a lack of control over the slave.[9] This moment of possible will in the slave, this uncertainty of power relations, is covered over through *giving* freedom—a narrative form. But this is also where there is a seemingly impossible radical point for departure, away from the narrative of emancipation—it is like a black hole, with a wormhole, and all we need is the mothership and we'll fly into other innumerable space-times.

In this same vein, this moment of putting to the ground—not as an anticipation of what follows nor as a continuation of what came before but as skip, as a leap, taken by Blackness outside of time and duration—becomes a doorway to worlds unknown, marks a crack in this antiblack world. When the ground (Black deaths) is taken from under the feet of sovereignty, as archive, as citizens, as police, as state, then emerges a need to cover this founding embarrassment.¹⁰ Such covering happens in Equiano's case through the slave master *giving* him freedom.¹¹ *Bodycam Exhibit 3* elaborates on how the archive and sovereignty—that is, citizenship and nations founded on the principle of one made out of many and many made for one, of texts deposited somewhere, authority given or taken to a place through the *contradistinction* to an unknown, immeasurable outside/blackness—are functions of antiblackness. That the listener engaging an archive, whether that be text or music, with hermeneutic authority, expressed in pitches' legibility and absolute countability, reintroduces this dichotomy, so does a notion of music as self-contained. But what happens in this work, within the center—the singularity in the middle that makes it have a hole—of this hold that is the musical work? There in the center, in the singularity of the black hole, is revealed another kind of story or telling, another kind of destiny (an "alter-destiny") than the one announced by the content, or the grammar that forms it and its overarching form, as the continual reforming of our ends and with such our stories.¹²

Musical pitches become related to the imitations and to the real scene, as always in musical sounding with/as social lives. Pitches become expression of Black lives, of lives lived in and with blackness. Imitation as not imitation but a sound of the real event, and as sound, it is not simply informed by the real scene but comes in from this moment as this moment (which is the end of a moment as self-containable). Of course, this does not imply that this moment becomes all-pervasive in a function of erasure of the moment, like the snare drum as antiblack paradigm. Rather, like the snare drum as spacing, the moment ceases to be localizable and becomes elaboration of space's out-of-placeness.

In the wake of Black Lives Matter protests following George Floyd's murder, the work of memorializing, through artworks, texts, and deeds cannot be left to falter under the authority of duration. "Wake work" is, as Christina Sharpe elaborates, the engagement of the antiblack world, as a way to celebrate Black life.¹³ To practice wake work is to theorize in black, to do work for, with, and from blackness.¹⁴ Wake work in black is to make Black lives matter. Rather than claim citizenship that does not attract bullets while knowing that there will always be those that "magnetize bullets," wake work for Black lives is work toward an alter-destiny, another worlding.¹⁵ As Charles M. Blow has uncovered, wake works for Black lives are being erased by duration (as a function of returning to the antiblack world).¹⁶ Wake work entails recognition of how Black lives matter is a matter of concern for this planet beyond an antisocial/antiblack world and thus must continue. This is wake work for Ben Mike Peters and for George Floyd; it is also a celebration of Black lives—here, there, and everywhere, beyond domains, to worlds unknown.

10

Black Life / Schwarz-Sein

Schwarz-Sein
Black lives' agency exceeds any accounts by antiblack methods of registering, identification, or control over narration. What Charles Uzor's works teach us is that there is a kind of agency that exceeds the delimitations of life, the properness of citizenship, speech, or belonging, as expressed by joining through opposing. It is those little holes in the fabric of space-time, when Black Swiss refuse identification, that sound another kind of listening, an interstitial listening. What is this interstitial (non)space that ends and begins space as singularity, like this leap in *Bodycam Exhibit 3*? There seems to be some kind of spooky action at a distance or wormhole-like connection, something like a mothership, that takes off, right there, in this interstitial listening, where we listen beyond antiblack brutalities to Black lives and their stories, their sounds, their music. What happens when we enter? Let's join this *Universal Consciousness* with Alice Coltrane, this music from beyond the veil, there where the ancestors dwell.[1]

When does the name of the citizen match the name of the nation? When is there an indisputable, absolute clarity that citizenship belongs to this citizen? We learned it from Uzor's *Bodycam Exhibit 3*: only in the moment, where Blackness is put to the ground, only in the moment where a noncitizen is identified based on their lack, because the other way around it will never work. Yes, there are identifications, like passports and ID cards or birth certificates, but these don't seem to be enough to make someone a proper citizen (albeit we might also say they are, in some ways, always already this same moment, if they are representations of it). There is no way to match the citizen with citizenship—to make, metaphorically speaking, the foot fit the impression in the ground.[2] Otherwise, the citizen would have to *be* the nation, but that is impossible, for no citizen can be the nation because *every* citizen must be the nation. This means that the nation is somewhere else, at some other place, as well as in each citizen, and this opens a gap between citizens and the nation. Like in the case of the sheep poster discussed in chapter 2, the "Sicherheit Schaffen" ad: there is some kind of unspeakable attribute that makes citizens into citizens that they carry with them always, that is in question in foreigners or Black lives.

Take the motto of Swissness for example: all for one, and one for all. If all citizens are for one and one is for all, then they must all be unique ones for this whole one, and the one must of course also be unique but, at the same time, in all these ones that make it. We may think this through with the archive, which is also to say, the scene of signs or names. Every instance of data or speech, every individual name, must be tied to the archive itself, to that which makes the instances have meaning—the one that all are for. But here emerges a gap, a problem; we might call this a singularity: the one has to be dividable without becoming less among the all, which is the whole. This means that no citizen can properly be the nation while also having to be the nation—both at the same time. But we might change our narrative here and recognize that this gap, this hole in the fabric of all, has been there before the distribution of it among its members. It had to have been there in the one already. Thus, the singularity, the black hole, is at the very center of the one, and thus also of each one and not only between them.[3]

What is the singularity? It is nothing other than the recognition that no one has ever been about themselves; all ones are sounds of what we aren't. We might put it this way: citizenship is not a pregiven thing; rather, it is radically open (or shall we say, silent), and this openness is worrying to citizenship. It is troubling because of the attempt to claim absolute and omnipotent authority. This is the same as saying that lives are radically changing all the time and that such is troubling the very meaning of what life is, if its definition is held in a moment in time or space. Thus, asking about the narrative structure, asking about the way things begin and end, is also asking about what they mean and ultimately about searching for a name. Our thinking through of the name is the unveiling that not only no name/term stands on its own but also that naming comes/returns/reflects nothing, emptiness, the unknown. Every term bespeaks an excess, or lack, in the relations of all terms that exist without it. Emptiness is not to be filled but to be empty so that life might flow, so that who is us may be written, as it has never been what we are but more than that, because it is who we haven't been yet and never were, and who we might become. This does not mean to separate the empty from its filling but rather to recognize that things *take over* emptiness. Thus, the empty emerges in the revealing of the colors of the world, in showing how things with names are always already only an image of this hole. The singularity is absolutely not one (by itself).

What does it mean to be no(t) one in the eyes of an antiblack world? What does it mean to be someone who embraces notness, in terms of double consciousness, or seeing the world through the eyes of others?[4] This, as we've learned, is a question of Blackness. While the world's eyes are producing whiteness as property against blackness, and thus the relation to this question of seeing is not the same for everyone, we have uncovered that there are different ways of losing your gaze. The falling over and out of a violent gaze that entails an opposition to others means entering a world where Black lives matter. And so to understand what such worlding makes a case for, I want to engage the way in which there is a kind of transmutation that Blackness bespeaks as it pertains to existence.

In his brief essay "Black Life / Schwarz-Sein," Alexander Weheliye pairs two terms that bespeak the Black experience: one from the anglophone world and the other from the germanophone. He has two reasons for pairing these two terms. The first is a philosophical question on the order of what it means to be, and the second is on the order of Blackness's refiguring of life and the world. As he explains it: "I use this term [Schwarz-Sein] both to inoculate the Heideggerian corpus with some much-needed melanin, and because it is the translation for the English language term *Blackness* as it circulates in German activist and academic circles."[5]

Weheliye places, or translates, "Schwarz-Sein" with "Black life." There is a proximity between Black Life and Schwarz-Sein that is expressed in the terms' uses as a protest, study, and political expression of Black people who move with blackness in a world where it is figured as the abject. Although one could translate Schwarz-Sein to "Black-Being," this translation somewhat misses the meaning of the term because of what the words *do*. If one translates Black Life into the German *schwarzes Leben*, then what gets lost is *Black* as nondescriptive because "schwarzes" is an adjective, but *Black* in Black Life is in excess of mere descriptor of a type of life (or skin color). In a similar way, Black Being can shift *Being* into a fixed noun, but in Schwarz-Sein, *Sein* plays out an invasion of the verb into the noun, a breakdown of the solidity of the noun as well as refusal of the separateability of nouns from verbs.

In this elaboration, alongside two political voicings, two terms, is revealed a transmutation of the scene of the political from a space of a plurality of enemies and friends, as ones in distinction to each other, to a notion of *the political as a staging ground for Black socialities*. As Weheliye elaborates, blackness as Schwarz-Sein modifies German philosopher Martin Heidegger's notion of "Dasein":

> In his philosophy of Being, Martin Heidegger deploys the term *Dasein*, which in German signifies a mode of being rather than subjectivity or individuality, and he does so with the purpose of not collapsing being and the human. Dasein denotes the sphere of being that establishes beings and things as entities. Still, Heidegger differentiates between man's capacity for world-forming, the animal's poverty in world (*Weltarm*), and the absolute worldlessness of the stone. For Heidegger, the tiered distinction between humans, animals, and minerals comes into existence primarily through the different ways these entities relate to death: "Death is the *own-most* possibility of Dasein. Being toward it discloses to Dasein its *ownmost* potentiality-of-being in which it is concerned about the being of Dasein absolutely.... The ownmost possibility is *nonrelational* (unbezügliche)." The nonrelational finality of death, or finitude in Heideggerian phrasing, creates self-reflexivity for the individual, revealing to them that they are a particular being and form a part of the ontological sphere of Dasein. The knowledge of one's own possible death leads to an authentic recognition of Dasein, which also constitutes the basis of Man's world-formingness. However, still lingering in the clearing of being-toward-death, Heidegger argues, "We do not experience the dying of others in a genuine sense; we are at best always just 'there' too." Even as Man's genuine encounter with the potentiality of his/her own mortality grants him/her the coveted entry into the exclusive Dasein club, the death of others, according to Heidegger, must remain stranded outside the establishment on a cold and rainy Friday night. Somebody must have forgotten to put them on the guest list.[6]

Black life, on the other hand, has a different relation to death: Blackness knows death not only as the finality of life but also as forms of life, because it is put into different forms of living death—death cultures, social death, and other necropolitical registers of death. Weheliye elaborates:

> I am left pondering, what if in all the hard work of always having to be somebody else, we cannot but experience relationally the death of others, because it is also our ownmost mortality? Put differently, what might all of this mean for Black Life's relationship to death or, in Saidiya Hartman's formulation, "the intimacy of our experience with the lives of the dead," be they in the form of humans, cultures, languages, genders, origins, belonging, and so on. In what sense does this familiarity with not the dead per se but the "lives of the dead"—and this is a very important distinction—supply an ontology of Schwarz-Sein?[7]

Weheliye points us in the direction of the habitual Be (also denoted as Be_2) within African American Vernacular English, as a way to explain the way in which *Sein* in Schwarz-Sein functions not as noun but as verb.[8] Be_2, Weheliye elaborates, "emphasizes the active and continual Being of Schwarz-Sein, the praxis of Black Life."[9] Thus, what Schwarz-Sein reveals is a mode of being that is with life rather than toward death, as also with, and as, the death. This Schwarz-Sein can be heard as being somebody else, as layers of somebody-elseness in *my* voice (in Weheliye's text, the multilayered voice of Michael Jackson provides an example for such elaboration).[10] This is a shift from life toward death, to life in poetic relations where we're always with and of somebody else, with and of what is not, with and of, as also for, what is silent.

In Schwarz-Sein is revealed that death is not an absolute measure for a position and standing in exchange relations, where birth and death certificates authorize life as proper, where papers delimit citizens as those who get to live. When death ceases to be about an end to a self-contained life, then endings become a care for lives beyond the end of one life. Black life is a radical living in excess of death as only an end of me. This does not mean that there is no *biological* demise, or death, but rather that thinking about death is revealed as a thinking about life and lives that cannot possibly have a beginning or end in death. Rather, this kind of life is a living image of those who've lived before and after and are not we, not here, but *still* us.

Thus, this is a question on the interstices between being and notness. As we've learned from Frantz Fanon, Blackness has no claim to being outside of being the other of whiteness (i.e., no claim to ontology). But as Calvin L.

Warren points out, Blackness also does not have any claim to nothingness because, we might say, claiming is a question of owning things (i.e., no claim to metaphysics). He puts it this way:

> The Negro is black because the Negro must assume the function of nothing in a metaphysical world. The world needs this labor. This obsession, however, also transforms into hatred, since nothing is incorrigible—it shatters ontological ground and security. Nothing terrifies metaphysics, and metaphysics attempts to dominate it by turning nothing into an object of knowledge, something it can dominate, analyze, calculate, and schematize. When I speak of function, I mean the projection of nothing's terror onto black(ness) as a strategy of metaphysics' will to power. How, then, does metaphysics dominate nothing? By objectifying nothing through the black Negro.[11]

But there is something that happens in Schwarz-Sein that troubles the absentness of nothingness and the presence of being. It seems that the very dividing line between being and nothing is made through antiblackness as the right to ownership. This also would mean that this border is, on the flip side, blackness as the in-touchness between nothing and whatever is.

In other words, the absentness of absence is reified in an overvaluation of the presence of presence as and in being. Schwarz-Sein is being without an absence to separate itself from—opposing absence to a present presence. Schwarz-Sein bespeaks an absentness within presence; it is a shift from standing into de-sedimented ground (a ground full of holes from these digs that Afrofuturists do). It is nothingness's presence in something, absences' hereness, or hereness as nothing other than a not-hereness. This is some kind of near-far confusion between nothingness and something. *There is a certain out-of-placeness to absences that we must think when absences appear in front of us*—maybe we must say nothingness is composing and improvising ([with]) itself) and comes as sounds of music.

Karen Barad poignantly points out the way in which, in the vacuum, a kind of nothing, being, and nonbeing are in a superposition that makes things nonetheless exist. Each bit of matter that eternally does not and does exist is made up of each other.[12] Barad elaborates:

> The explanation physicists give is that the lone ("bare") point particle's contribution is infinite as well (infinitely negative due to the negative charge of the electron), and when the two infinities (that of the bare electron and that of the vacuum self-energy) are properly added to-

gether, the sum is a finite number, and not just any finite number but the one that matches the empirical value of the mass of the electron! In other words, an electron is not just "itself" but includes a "cloud" of an indeterminate number of virtual particles. All this may seem like a far-fetched story, but it turns out that vacuum fluctuations have direct measurable consequences (e.g., Lamb shift, Casimir effect, the anomalous magnetic moment of the electron).

So even the smallest bits of matter are an enormous multitude. Each "individual" is made up of all possible histories of virtual intra-actions with all Others. Indeterminacy is an un/doing of identity that unsettles the very foundations of non/being. Together with Derrida we might then say: "identity . . . can only affirm itself as identity to itself by opening itself to the hospitality of a difference from itself or of a difference with itself. Condition of the self, such a difference *from* and *with* itself would then be its very thing . . . the stranger at home."[13]

While this passage gives us a beautiful account of how nothingness, how our notness, is us, we must still problematize this scene. It could easily be conceived as a conflation of notness and being, where any account of notness ceases to be possible. The conflation of notness and being can lead to an erasure of notness as the claim to having, owning notness, just like Warren elaborates that nothingness becomes an object held (in opposition/possession) by a certain register of study.

In some ways, this is about recognizing that "being" claims to be, or that nations and citizens claim to be, and that such is in opposition of notness. That is also to say that being claims notness—that notness is merely an effect of it, that notness is nothing other than it. Then notness becomes *turned* into being, is owned by being. It's like the discussion of silence and sound in chapter 7, with John Cage's *4′33″* and Uzor's *8′46″ George Floyd in Memoriam*. How *8′46″ George Floyd in Memoriam* presents a critique of a kind of possible erasure of silence as a claim to being in control over silence, even when such is bespoken under the heading of absolute lack of control. What is proposed here, in Schwarz-Sein, is that being does not "include" notness; rather, notness *includes* being. Or, more accurately, and to reference a brilliant work by the Sun Ra Arkestra: *Nothing Is*.[14]

Any account that does not allow us to think silence, or notness or that without power, through its insistence on finding presence, being, life, or existence, everywhere, can run the danger of erasing notness, or silences, or those who are no one. This is the erasure of the gap between nothing and something, which is the erasure of the border zone. With such it may turn into the erasure of the interstitial that is found in the underground of borders as the excess of what is. It is very much a question analogous to the question of racism and race in Switzerland, or in any color-blind space: how it is not enough to merely claim that the here, this presence, is already diverse, because such is also color-blindness. Rather, we must problematize the very erasure of how hereness claims itself to be here and does so by erasing absences, erasing its end, into borders or nonexistent spaces, that are interstitial and black.

At the same time, a similar refiguring as Cage's *4'33"* along the lines of Uzor's transmutative *8'46" George Floyd in Memoriam* can bring to the fore a black radicalism.[15] Rather than reifying the one, the individual, the self, as being itself, by way of erasing silence/nothingness, a reversal of such celebrates notness's exceeding of what is and what is as an expression of what is not. A reifying of appearances as owning nothingness through holding it but being separate from it, and a positivist world where everything is transparent to each other, can lead to antiblackness as the gaze that sees nothing always as its opposition. Thus, rather than erasing silences and nothingness with positivist agents and sounds, all such appearances must be thought out from the unthought that blackness marks. Or in other words, appearances, because they appear, may fail to see themselves as simply/mere appearances (of nothing). They might claim their worth from being entangled, they might claim their worth from being not nothing, but they are not. There is no separation between their notness and their appearance, and such is not because they are but because *they are not*.

Black Life

This reversal from agents and reals to imaginary unborns is unknown thought-work given to us by Uzor's musical compositions, elongated here into the cosmic theory of general hesitancy/relativity.[16] It is a sociological Du Boissean thought-work from the unknown. Blackness is appearances' breakdown as self-contained; it is beyond identity as a question of (self) relations. Active from 1972 to 1980, the Black feminist organization the Combahee River Collective's seminal statement is a celebration of such Black radical practice:[17]

> Above all else, our politics initially sprang from the shared belief that Black women are inherently valuable, that our liberation is a necessity not as an adjunct to somebody else's but because of our need as human

persons for autonomy. This may seem so obvious as to sound simplistic, but it is apparent that no other ostensibly progressive movement has ever considered our specific oppression as a priority or worked seriously for the ending of that oppression. Merely naming the pejorative stereotypes attributed to Black women (e.g., mammy, matriarch, Sapphire, whore, bulldagger), let alone cataloguing the cruel, often murderous, treatment we receive, indicates how little value has been placed upon our lives during four centuries of bondage in the Western hemisphere. We realize that the only people who care enough about us to work consistently for our liberation are us. Our politics evolve from a healthy love for ourselves, our sisters and our community which allows us to continue our struggle and work.[18]

Something of the individual breaks down in the Combahee River Collective's use of identity. It is about a care for us. Identity that is an appearance is claimed as care for each other from an unnameable position that refuses a more truthful backdrop than our appearance, which is (valued as) nothing.

Antiblackness seems to work as a methodology that erases identity, as well as that behind the marks (names) of identity, which has no value (and creates the separation into name and abstraction). Scholar Keeanga-Yamahtta Taylor explains:

> But "identity politics" was not just about who you were; it was also about what you could do to confront the oppression you were facing. Or, as Black women had argued within the broader feminist movement: "the personal is political." This slogan was not just about "lifestyle" issues, as it came to be popularly understood, rather it was initially about how the experiences within the lives of Black women shaped their political outlook. The experiences of oppression, humiliations, and the indignities created by poverty, racism, and sexism opened Black women up to the possibility of radical and revolutionary politics. This is, perhaps, why Black feminists identified reproductive justice as a priority, from abortion rights to ending the sterilization practices that were common in gynecological medicine when it came to treating working-class Black and Puerto Rican women in the United States, including Puerto Rico.[19]

Antiblackness takes away the care for us, for Black lives, for Black women, and turns identity into something that only bespeaks the individual. The Combahee River Collective is critiquing an antiblack world that has always already closed the question of *who us is*. In the affirmation of appearances by those who are no-one, which is also to refuse identification as a method to ownership because they disturb its valuation, lies a breakdown of the separation between name and abstraction, between value and living matter, between I and us. And it is the unname of blackness that marks the (un)naming of the world as a refiguring of values, as a celebration of appearance, as a mereness that reveals the nothingness of appearance and frees nothingness from the shackles of being less worth than relations—of appearance being only worth its value as name from exchange relation. All that is is really just nothingness's music.

Nothingness is the undercommon of things in relations that are poetic (opaque)—things that are not only in relation but also where relations form. The notion of an absolute nothing, one that is the opposite of relations, as the exception that provides the ground for relational exchange value leads to another Derridean scene of substitutions, a search for the real thing. This is revealed in Jared Sexton's essay "Basic Black" without spelling it out as such. In it he describes how today there is this search for ever blacker black colors, and that such is a question of economic gain, because whoever makes and owns the blackest black makes the most profit. What his essay shows is that there emerges a search for a more "basic" blackness, which separates blackness, or nothingness, absolutely from what is, and that such search is basis for contractual relations in the form of selling ever more black colors and artworks, thus also bespeaking a world of citizens who are because they can hold/own nothingness, blackness.[20] Thus, if nothingness and blackness are kept at a distance—whether that be through approaching it, through holding it, through being separate from it (i.e., not *with it*)—then nothingness and/or blackness become continually displaced into further absence from here.

King-Ho Leung's elaboration of nothingness in Moten's work, in comparison with three giants of European philosophy, is telling here.[21] There is a certain problematic that is revealed in the contradistinction of nothing and being that is of the order of the qualifier *absolute* as positioned in contradistinction to relations. Leung points out:

> What we find in Moten's paraontology is thus a reversal of the traditional metaphysical or even onto-theological privileging of Being over nothingness, where nothingness becomes the center or even "ground" of everything. Being only "is" by virtue of being in an antithetic relation with absolute nothingness; it only exists as "anti-nothingness" (as anti-blackness) or what Schelling calls "not non-being" (*nicht nicht Seyenden*). In other words, Being as *anti*-blackness is always merely relative,

whereas blackness qua nothingness is *absolute*: as if echoing Schelling's speculative notion of the Absolute, blackness is what Moten calls "the absolute, or absolute nothingness."[22]

I want to take a step back, with Leung to Moten, even before the before of being, to where antiblackness is not being, where being is not antiblackness. That is to say that in excess of antiblackness, and *absolute* nothingness, is a nothingness that is articulated in (excess of) relations but with an attention to its Schwarz-Sein. If nothingness is all the things that are, as we've elaborated with Barad, then *nothingness is the recognition of appearances' appearance, and that this doubling of gazing, this double consciousness, is a poetic relation seen as a clarity of seeing the opacity of relations*. We hear here nothingness in excess of an absolute as the opposite of relation, where such nothingness becomes more than absolutes and relations; it is that interstitial space out of which they both come and distance themselves from once they mark such as a border.

There is a strange hyperpresence of appearance that is also its failure, that the white gaze locates in blackness, that an antiblack gaze locates in nothingness. In the breakdown of the law of holdable appearances that blackness marks is the opening of lives beyond control by ends. Marriott helps us understand this via "corpsing"—the failure to perform/act a role:

> Corpsing, therefore, seems to have been defined more precisely as the death of theatrical artifice. But corpsing is also evident outside theater; we see it when people fail to live up to or grasp their social roles. Hence the derisive laughter attached to those who forget themselves, or have their pretenses exposed, or fail to convince us of their authority. The promise of a role is meant to accord with the performance of desire, and corpsing occurs when desire violates or threatens that promise. In this way the codes of social performance are used to discipline desire like a bridle, insofar as one's persona is taken to be more than a formal tie of social being. This is why the essence of corpsing is the violation of rules of prescribed performance under the command of social laws, and consequently those who obey the rules are said to be at one with their roles and not regarded as subjects of them. But what if one's role is to be socially that of failure or if one is ordered and commanded to perform a role through one's corpse-like obliteration, would this not mean that corpsing can only occur when one refuses that spur and its contagious pleasure? Would this not be an example of a "death" of death, so to speak?[23]

Isn't the marking as failure the cover-up of the performance that is before and outside of the performance? That is to say that when the performance is seen as failed in an exception to the rule, then such ascription of failure covers a rupture that reveals performance itself.

We also remember the similarity in form with a certain idea of transgressing death in necropoliticality that requires the killing of death, of no-bodies, of Black lives, because through such is claimed authority over death. But in Schwarz-Sein, death cannot be an event nor a nonevent as exception of events but must be with living. Furthermore, not only is the performance of proper sociality also attempting to erase and control death but it is also producing a proper performance through its relation to an exceptional mereness of other performances that are failures of proper performance—they're mere performances as corpsing. Thus, failure is the name, like corpsing, of that performance that is not the authorized proper performance. By marking and thus owning the ascription of corpsing, the properness of one performance is established and is utilized as an erasure of its own performance as performed—it erases its own temporariness, its being as a performance, and makes itself into a more fundamental real/truth.

It is here that we may follow the interstitial and think such performative appearances not as pregiven there but as *performances*. Édouard Glissant sort of points to this in an unfinished text where he nonetheless also departs from this kind of elaboration. At the same time, it is unknowable how we must read this text. Thus, I read it anew, this draft, this sketch, this blueprint. He writes:

> If we suppose that nothing is true, we overstep continuity, apart from the continuity of concrete things which have no need of capital letters. For Truth, capitalized, does not go through anything, the Absolute isn't anything. And the Absolute is True only insofar as it surpasses the absolute. Thus there is no continuity from truth to truth, and any continuity is in the falsity of things, which we have to fight. But living is continuity itself which, if it ceases, gets into rest—what we call death—to prepare another continuity. Thus, the cessation of continuity in the language of who claims truth is a decisive end, which leaves room only for useful truths. But the end of continuity of the living questions us endlessly, not on useful things, but on unexpected passages. The continuity of Truth does not loop in any way, it is a fragile line of fire. The continuity of the living is a spiral, which is not afraid to break.

Rupture

> And it's not only the living that is not afraid to stop, but it seems that the rupture is one of the steps of its advance. The rupture of the living is often the chance that is in it and that builds it, unmaking it for an elsewhere or an otherwise. Chance is rupture and at the same time continuity in the living, without the need of dialectics.[24]

Reading Glissant's rupture with Marriott's corpsing allows us to think failure as rupture's exceeding of contexts, which is what allows for an-other chance, makes such failure the very excess of a death-world, of absolutes or relations, relations of absolutes measured as ends. Thus, what the second sightedness of Blacks/blackness does is reveal time, as beginnings and ends, as ended: blackness is outside of time as measures of continuity, as relations of death-truths, and it makes time as more than self-contained moments.

What our celebration of appearances as appearance does is reveal their notness, reveal that performance is nonperformance as performance's performance. Hence, if Blackness is the very condition of excessive performance as corpsing, then regardless of whether such is refused or affirmed (by Blackness), Blackness is excessively performing, a performance that is uncontainable as mereness for/or truth. If corpsing is failure and is marked as blackness, then such failure is not of blackness but of the order that performs and claims such performance as truth and all other performances as corpsing, as mere performances, as exceptions. There is one proper way of being and all others are mere appearances, mere identities, and not this real thing that is proper life. Blackness can perform and fail (to play truthfully), or fail as failure (perform wrongly as a *mere* performing or as corpsing); it doesn't matter because either way such failure must be located somewhere (in black), and by locating, it is made into an exceptional event so as to claim the authority of the one performance over others, effectively making such performance into the real/truth/proper. Consequently, if Blacks fail in performing socially, such failure is already covered over by the white gaze through the exceptional event of mere failure, of mere wrongful, improper performance and marks another performance as more than mere performance.[25] Although, it is important to note here that the resistance to failure as an attempt to undo the doubling of the white gaze, if done as a claim to authority over and of one performance, is antiblackness, is the ruse of performance's truthness—or an idea of a truth behind performance that then is some kind of raising of performance to truth. Thus, an evasion, or erasure, of the fallibility inherent in performance functions through a localizing of corpsing outside of proper performance, functions as the deputization against failures, against Blackness.

Furthermore, Blackness must not be found anywhere, must be absolutely absent. The erasure of blackness from here is thus the evasion of performance as nontruth. Which is also to say that the presence of blackness, of absence, is the revealing of how there is a performance/appearance that claims to be proper, to be it all, to be real. This is why Blackness from abroad can reify antiblackness in European contexts, or why blackness can be displaced onto those who are not men, or those who are not women, or those who have no heritage to claim—these legitimize the erasure of blackness as here because it is absolutely found only outside of here, of proper life, of presence. The second-to-last point is of concern in the Black Swiss context where sexism is acknowledged but racism isn't, and thus *the illusion of a nonracist society can be purported by an inclusion of Black lives on the basis of gender while erasing their Blackness.* This passage by Campt is quite telling:

> In contemporary Germany, this eurocentric notion of beauty rejects women of color as "other," while paradoxically giving positive value to "otherness" as "exotic," and thus exterior to this ideal. These Afro-German women experience the adjective "sch[ö]n" [beautiful] both in terms of a rigid European beauty standard which excludes many ethnicities and as a category which *includes* them, albeit as marginal or extrinsic in relation to this standard.
>
> Moreover, the "exotic beauty" of Afro-German women is also valued over that of African or other darker-skinned women of color, based on their lighter skin color. For their lighter color is perceived to be "not as foreign" as the "blackness" or darker color of other ethnic groups.[26]

Blackness becomes erased through their proper genderedness (they are "beautiful" women), and antiblackness is erased through the proxy of the "exotic beauty" (that is, through making them women who may be allowed to be admired). This is a double erasure of antiblackness, for by partially including them in society, antiblackness as well as blackness are relegated to the outside of this European, or national, space.

A Swiss example is given by Tilo Frey, the first woman, and Black woman, in the Swiss parliament who was criticized for her dress choice in 1971. She wore a white dress while all other men wore black and the white women wore black with colorful accessories (see figure 10.1). This must be understood in relation to how the black suit "became established as bourgeois men's clothing in the 19th century [and] . . . associated with a masculine ideal of politics. . . .

In contrast, fashion, pattern and color became synonymous with femininity," as Jovita dos Santos Pinto explains.[27] The newspaper found her dress choice unfit, separating her from both women and men. But in fact, the first cantonal council woman Raymonde Schweizer also wore a white dress in 1960, but her choice was deemed appropriate for a woman. In the controversy caused by her dress choice, sexism and racism met, but race was doubly erased: as a question of proper dress choice, it became a question of appropriate clothing entangled in gender, but the issue only appeared because of her skin color, which, of course, was never directly addressed. Thus, rather than seeing this as two forms of discrimination, this scene actually reveals that Blackness is covered by discourses that reduce this scene to misogyny as well as its critique. Dos Santos Pinto problematizes the idea that Frey was an "example of a double emancipation."[28] As she points out, race completely disappeared from the discourses and from historical consciousness (like Frey herself).[29] But I would go further and say that in this scene of emancipation, antiblackness is erased on the basis of a claimed success of emancipating and including women. Firstly, Frey's blackness is erased and then Frey herself (thus there is also a relation between the erasure of women in general and the absolute need to erase blackness).[30]

Similarly, men can become partially included through gender as well, through joining a societally accepted form of maleness that is equated with citizenship. This is how Hans Hauck was able to join the Nazis, and how they recognized Black males—as from Black countries, to have relations with, but that then ultimately also, because of their existence, can be proof of a hierarchy of races.[31]

This structuring paradigm also plays out in the way in which discourses can be held about Black lives. The historical elaboration of Black studies in Germany has, as some have pointed out, had "a blind spot" for Black men.[32] I would contend this means Black lives in excess of proper gendered inclusion even as "lack." In other words, the critique genders by assuming that such lack regards men, but as unthought these lives are ungendered—i.e., their gender identity is not a priori given. (They can be men, queer, etc. and also ways of gendering that cannot be thought, since gendering requires already their inclusion.) As Rinaldo Walcott points out, regarding Black studies and queerness in an essay published in 2005 but still relevant today, the black study project's liberatory potential requires thinking with the unthought.[33] "Black" exceeds any delimitation already given by a world defined and borders policed on the basis of the destruction of blackness.

The various tactics of inclusion aim to erase the possibility of blackness, and of Black lives, exceeding the confines, the delimitations necessary, for an antiblack world where value and resources, capacity itself, is always routed in one

Figure 10.1. Tilo Frey in the National Council as printed in *Feuilles d'Avis de Neuchâtel* on November 11, 1971.

(proper) direction. Lastly, this also shows that proper (social) performance not only requires the performance of failures—that is, the performance of blackness or *minstrelsy*, which is also the claim of an antiblack world to an authentic form of blackness—but in fact emerges in the control over the failure/ends/borders of performance. In fact, the proper performance emerges in the failure of performance—we might say, the control over nonperformance and its boundaries. But as we've uncovered with *Betty's Case* (chapter 6), when blackness, or *nonperformance*, becomes "Black"[34]—the very uncontainability of nonperformance—then categories of identification of subjects, of beings, on the order of their *unremarkable* performance, undo properness's existence in contradistinction.

The incapacity of Blackness (to perform) that always accompanies Black lives, that non-Black life can *seemingly* successfully erase, is what reveals the way in which there is simply performance and that such is the failure of performance as all and everything.[35] In other words, (social and other forms of) performance as a nonperformance and as a site of possible failure shows us that we're performing and that there is never a complete performance—that is, a whole system of values that remains as a measure of it all. It is in Blackness, this uppercase, as an engagement of antiblackness as not all there is, that demands us to exceed the world as one moment or event, because it is about a togetherness that exceeds meetings as exceptions/events, as things that are.

This is also to say that the context that allows for inclusions can only be refused and undone in a critique that flows beyond critique because critique is always tied to the contexts of the world. In the excess of critique is a kind of celebration of blackness, of that which is always reminding us of the excess of and remainder of one performance. It is the celebration of the way in which critique must always already be able to hear the world as ending, the context as changing—the end is before the beginning, we might say. That means, in a general sense, that in the marking of a failure as an event, named *corpsing*, lies the illusion of the evasion of failure, and such becomes antiblack because failure must be prescribed onto someone else, must be locatable somewhere else, and is always close to and of blackness in this equation of modernity.

The white gaze attempts to cover the continual failure of performance, tries to cover the appearance of society, which means also marking what it is not as its failure, its problem or negative other. That is the claim to omnipotence, to authority over space and time and all there is. In a doubling down comes to the fore blackness as that which flows out from (and way before) the failure; failure becomes an endless falling. It is to fall into the black hole as entrance to the mothership—or, like a J Dilla beat, it is to "Fall in Love" with . . .[36] It is here that failure falls over into a celebration as an unbounded love, as a motherly celebration—"fall into the world of things" is a motherly practice of care beyond and in excess of proper relations.[37] *Incapacity* (of appearance) as failure of proper appearances that is blackness is the very unrooting of capacity as the border and staking claim to be authorized and authenticated in the uniqueness of one against none. Blackness is the magical unperforming of the world that is the incipit for performance and nonperformance of worlds. Thus, we hear-see how such is the impossible improvisation, as musical rupture, that is the composing of the sound of a better world.[38]

If blackness and nothingness can be known, or held in finite points, then they simply keep shifting away (over time, over space, over lives, and all other possible directions). But from what? Nothingness shifts away from nothing/blackness. Nothingness that is measured, no-bodies that are controlled by value, is not speaking of nothingness that is valued but exchange value itself—contractual relations. Blackness must be open to change, is the very description of presence's ever-changingness, as the continual making of new performances. This is why antiblackness terrorizes Black lives, Black knowledge, Black history, for this brutal imposition is what antiblackness needs to be all-encompassing authority.[39] Studying Blackness amounts to opening ourselves, our world, libraries, spaces, schools, and more to the fact that they cannot remain without us, that they need to be lived, read, studied, and thus changed. These texts from beyond the veil, as the beyond's already-hereness, provide us with an *upsetting* of views (yes, we dub with the upsetter Lee "Scratch" Perry here in Switzerland) through improvisations of Black life / Schwarz-Sein.

For Nahum Dimitri Chandler, W. E. B. Du Bois's theorizing proposes a similar blurring of presence and absence. How his study of African American lives was a superposition of the specific, or local, and the global, or general.[40] It is about the role of the example, or the case, the case as study, the example as incipit for thought. Chandler explains: "The crucial sense for contemporary thought to reckon with is that, for Du Bois, the Negro American example takes its place as the incipit for him as a thinker, as fate or instituted chance, overdetermined in both its freedom and its necessity. The example given to him, that which solicited him most immediately, ineluctably, and without ceasing, posed a question about possibility—supposed as ontological and historical, *ontohistorical* in the traditional senses of ontology—that yet remained exorbitant for traditional formulations of philosophical question in the modern epoch."[41]

Thus, the example is, in fact, that which opens thought rather than closes it; it is that which is before the question, before the stating of the problem as it shifts thoughts that move as contexts (or states, like states of mind, or nations). This is to say that Chandler elaborates for us how the example of a thinker, a text, and a concept, as a problematic, that is something unthinkable, opens possibilities for thought. Furthermore, thought is given not as relation in contradistinction to the unthought but as unthought relations: these thoughts are of unthought contexts. That is to say that it is not thoughts as, or in, relations that make new thoughts, or sounds, or sense, but unthought relation, that which relation leaves unthought, that which is relayed by unthoughtness. Poetics of relations are made in the nothingness of the abyss of the slave (on the mother)ship; that is what Glissant taught us.[42] This is unforeseeable, impossible improvisational generativity—it is blackness.[43] The reason why it is announced in the given but "always other than simply given" is because it is in the shift, in the doubling of the white gaze, like in *Bodycam Exhibit 3*, that it becomes obvious that appearance is nothing other than nothing.[44]

The example as case has always been about its emptiness—that is, that which bespeaks its changing, both as in what may be inside and what is possible as case.[45] The existence of the case—its appearance as a kind of presence that is it being brought to judgment—is in opposition to its notness.[46] But its worth stems from the extraction of nothingness—of what it is not, from the world—that it announces. A case that cannot bring anything is worthless—like the case of antiblackness, which is the situation of living as Black Swiss, cannot be brought to court, but the white gaze, which is the owning of blackness, can, and its case is not only heard but becomes affirmed when the gazed-upon is deemed too black (i.e., is supposed to show identification or be expunged from the inside). Antiblackness continually figures all existence on the basis of its ability to bring in notness, to extract—which is also to say, control—blackness (i.e., its end, or death, or outside). This is a scene of a certain contractual basis for sociality if the social is that space that a nation (or a group of subjects) circumscribes with control/authority. And as such it is also a question of the way in which the inside, or that which is, exercises control over the outside. In the metaphor of the jug as case (like a casing or vessel), the inside may be filled, thus its emptiness is taken into the structuring of the case through the control of such emptiness by way of knowing what may be able to be in there. In politics, the voices of those who are properly registerable as citizens similarly must not pose a threat, and are thus an always controlled unknown. My point is this: the emptiness of the case is always in excess of the case as case, even when seemingly contained by this case. In fact, the improvisation of how we make jugs, which may change a jug into a black square that can't hold anything, is the danger that the existence of the (proper) jug aims to avert. This means that unthought voices must come before the authority that bestows some the right to have a sound.

If blackness bespeaks the notness of being, of life, and thus that which antiblackness opposes, owns, approaches, or contains, then black equals death.[47] But what kind of death, or what is dying? Antiblackness is the claim to absolute authority over how uncertainty, how the outside, how death, may enter: it is relations controlled. Antiblackness's base condition—that is, how it may emerge—is the claim to control how life ends, thus it can define what may count as life. But isn't it clear that all ends exceed their meaning, which is their name, that is also the claim over relation as the authority over what is and what is not? For what is announced in endings is something unknown, unheard, unthought. Antiblackness requires the claim to never having had an ending (or, we might say, it must *have* an end, as in it controls and owns its ends). That is a possessive relation to ends, which is using death as a negative other for life. But circumscribed with death is in this scene all that changes what it means to be alive, what life is in general. While antiblackness appears to be the fabric of the world, Black life / Schwarz-Sein bespeaks the breakdown of such claim. Black study is about the uncertainty that knowledge always bespeaks.

The interstitial space that is not, that is nothing, is that which (un)founds more than all—for emptiness bespeaks anything that could have ever (not) been. Thus, interstitial listening is to study the world and what it could have been, what it was, what it is, and what it might become. This is black study, and it *has* no end (in productions of knowledge as property or ownership); it has always already ended and is after the end of the world. Black study is a continual listening and reading like Jasmine Farah Griffin's writing practices outlined under the heading of *Read Until You Understand*, which hints at an understanding that never comes to an end;[48] we must keep reading. We must write and speak, study and sing, with, as (and) for, Blackness, like Du Bois's city directory. To bring the failure to claims over things is the celebration of things and their nothingness, the celebration of blackness as the doubling down of critique where both the world is critiqued and its context too—the failure as context-dependent becomes unchained and spacing for new worlds (in all directions of space and time).[49] Chandler's elaboration of how Du Bois is example and is incipit is thus not to be read in the direction of an entrance into an order of things present but rather as the very underground before and outside of any event, where any event is always of the ground's grounding—a *grounation*.[50]

In this limit of the world that is the limitlessness of worlding as black study/work lies another kind of politics that, as Black life / Schwarz-Sein declare, is the hesitant study of lives in poetic/unthinkable relations—that is, our association and identification across borders, in excess of us as us. In this is outlined a political project that is before politics, as before and after but also as petitioner with*out* standing, that improvises political insurgence.

Conclusion
Alongside a Chorus of Voices

Dreaming of Future Festivities

It is through our musical soundings, like reading these notes on the page, that another world may become audible. To conclude this book, I want to share with you a piece of music that I wrote, that reflects the concern of this book, and made me hear new possibilities in the world. My work, *Alongside a Chorus of Voices*, was commissioned by the Lucerne Festival with support by Pro Helvetia for its newly formed festival focusing on current music, Lucerne Festival Forward. It received its premiere performance on November 20, 2021, with musicians Emily Brausa, Frauke Elsen, Kevin Fairbairn, Sam Jones, Edward Kass, Marina Kifferstein, Benjamin Mitchell, Ben Roidl-Ward, Noah Rosen, Ona Ramos Tintó, Wendy Vo Cong Tri, and Andrew Zhou, conducted by Mariano Chiacchiarini.[1] Led by Mark Sattler and Felix Heri, a cohort of younger musicians from all over the world cocurated the festival together. One can only wonder as to what the statistics are of previous occurrences of Black composers commissioned, or performed, at the seminal Lucerne Festival, and even more speculation would be necessary to inquire when a Black

Swiss composer was featured. (Following this new offshoot of the main festival, in the summer of 2022, the main festival made "diversity" its theme and featured numerous Black musicians and composers, mainly from the United States and United Kingdom.)²

I see a possibility in Lucerne Festival, announced by Lucerne Festival Forward's underground, announced by the people working behind the scenes of a big institution enmeshed in an antiblack world, a possibility for Black lives to speak and sound an impossible future, an Afrofuturist incursion, toward a reshaping of our world—music as a force that shifts the grammar of even those structures out of which it can emerge. I want to talk about this experience, about the way it came about, about the music, about the spaces, about what had been said, heard, or ignored. I want to do this here because it is like this book; it is of a similar kind, it goes alongside this text, and the writing of this book goes alongside writing music. It goes alongside meeting Charles Uzor and the experience of Black Lives Matter, and so many more.

When beginning the work on *Alongside a Chorus of Voices*, it was apparent to me what contexts I was writing it in. It was in engaging the contexts elaborated in this book of Blackness in Switzerland that I composed in a manner that I hoped to be a possible opening to talk about the unthought, and the only way to do such work was to be fooled already (easy, as I was already an *unthought* coming into appearance), like the future was already there in that present. Thus, I begin with pencil and paper, with my own life story, with all knowledge of these fields of power and life—the ways in which antiblackness and blackness can both be found here in these spaces—and with the knowledge of how the offer to write came to me. This offering disguised as an offer for business came from the conditions of this new musical space within another. As mentioned previously, the curation of the festival happened through a collective of younger musicians who were alumni of the Lucerne Festival Academy at the time—a program for musicians up to thirty years of age to study and participate at the Lucerne Festival. Some of the musicians of this consortium were collaborators of mine, and two of them suggested that I be commissioned for the festival. Johnna Wu's advocacy in particular, among others, such as Marina Kifferstein, convinced the team to commission me as the first commissioned composer of this exciting new festival. Being aware of the context of the festival and, most importantly, the (non)relations between Blackness and Swissness, as well as the "myth of absence" and the lack of awareness by curators about Black (Swiss) composers, the work was already positioned as a kind of opener for thought and lives.³

As mentioned in the introduction to this book, the year before this festival, in 2020, I was part of a conference organized by one of the most preeminent ensembles of new music in Europe, Ensemble Modern, as well as my mentor, the brilliant George E. Lewis, discussing this myth under the heading of "Afro-modernism in Contemporary Music."[4] This conference discussed the ways in which Black lives are excluded from narratives, institutions, curatorial decisions, and funding under the banner of new music. This is not to say that there aren't numerous Black musicians who were part of the history of new music and who traverse it today or whose work could be considered relevant to and for new music. Rather, new music as a field never confronted the antiblackness of the structures within which it finds itself and which it, because of this, further perpetuates. This conference, and the following and antecedent work by Lewis and others and the globality of the Black Lives Matter protests, opened new possibilities of discussion and thought in Europe and the Americas for a critical re-problematization of questions of antiblackness and blackness, especially within fields and spaces that had assumed to have transcended race.[5]

In *Alongside a Chorus of Voices*, I engage the questions of Black lives and Swissness through the use of bells. Bells have myriad associations, meanings, functions, and material/sonic qualities. They are a symbol and sound of Switzerland: the sound of cowbells, used to locate cattle, is a very common soundscape within the geographic space of Switzerland. Bells are also a national symbol for Switzerland. Furthermore, bells used as organizers of time and space, used to localize things and to announce things, are also entangled in modernity's freedom-slavery axis. Used to announce the emancipation of slaves and other moments of freedom (such as, for example, the famous Liberty Bell), bells were at the same time also used to locate slaves, to keep them from being able to run away.[6] Thus, at once bells signify freedom and slavery as well as pose questions about our relation to space. Freedom and slavery are, as we've elaborated, both formed within the scene of antiblack slavery, and freedom is a continuation of antiblackness and not its end, although I hear it as its ongoing ending, as an abolitionist working over of endings. Lastly, bells also have an altogether different aspect to them; namely, bells have a complex sound—they are not one in their sounding. A bell sound has multiple competing fundamentals or prominent frequencies—that is, musical pitches that can be perceived. With some bells, different people will identify different base frequencies; in other bells, there is a transformation of the base frequencies over time, bells might form chords (with a variety of intervals), and others can produce sounds that appear only in people's heads—so-called summation and difference tones. The last aspect is, of course, also related to the space and to the positioning of the listener within that space as well as all the other things in the space that reflect sound. Thus, the sound of many bells is largely immeasurable and somewhat unpredictable and uncontrollable.

In my piece, all these aspects of bells are put to use. Bells are first heard only as markers of movement—flute, oboe, and clarinet begin among the audience and walk around in the audience during most of the piece, wearing small bells that ring when they move. Bells are then suddenly *heard above the audience's heads*—like stars, these guide the walking musicians. At the same time, these bells also represent the sound of freedom in two senses. On the one hand, in some of the moments with bells ringing on the ceiling, they are like a promise for freedom in an analogy to freedom bells and mythical stars used as guides into the North/freedom (in the history of slave emancipation of the United States). But, on the other hand, there is an alternate way of hearing these bells; namely, they also confuse the listener, those that try to follow the movement of the marked musicians, making it more difficult to localize them (does following stars undercut state-sanctioned measurements of life?). Slowly, the bells multiply, and it becomes clear that the emerging music itself is coming from this complexity of sound (or we could also say, of listening) that the bells have because their frequencies are opaque and relational (or complex). Thus, the work unfolds into and as music—but only when listening becomes removed from authority, from measuring and tracking. The work starts moving around every audience member and musician in unique ways that *are*, nonetheless, always only through relations.

This use of bells points to an engagement with the question of Swissness, blackness, *Schwarz-Sein*, listening, and sovereignty, which uncovers both antiblackness but also another story. In the interstitial listening that this music needs, a poetics of relation comes to the fore. What happens here to citizenship in this superposition of blackness and Swissness? Beyond the question opened by this music, by this book, by this name, that is the meeting of lives. In *Black Swiss* lies a radical refiguring of what belonging means, what living in space means, what moving in time means. It is like these bells: it is about how bells need to be acknowledged and, consequently, they cease to be absolutely locatable. As a result, they speak of how the protest that is *Black Swiss* turns into a celebration of lives in excess of relations and identificatory violence to erase our sound. The sound of music, of us, only comes from this black study project, from this giving up of absolute positions, of the revealing of the bells that make space, that become only bells through music in the first place.

Toward an Alter-Destiny

This process of thinking and doing in black, which is also always a thinking and living for and with others—of giving to others the space, time, resources in general for life and doing such over time as well as space—is the political (non)*power* of blackness. By power I mean here something like *A Power Stronger Than Itself*: a book documenting the Association for the Advancement of Creative Musicians (AACM), where making music in common with living lives flows over into all domains and into spaces unthought.[7] Thus, rather than ascribing any modality, domain, or category to blackness, which always fails as it is again a strain toward authority, blackness is that from which any act comes from, not as agency but as something else beyond (un)nameability. Our elaborations point to the way in which Black/blackness is a kind of working, a practice, something like improvising and composing new sounds, maybe learning to listen, in excess of a doer or any given notion of what may work. Blackness "cuts the distinction between essence and instance."[8] Black, thus, is at once about Black lives and about more than life, for it is also in touch with the death and with the future not-yet-here. Blackness is the very refiguring of what us may be.

The refusal of antiblackness is spoken as a care for Black lives—Black lives matter—as a black study project that is living (with/in/as) life (in Black). *It is not that there is some more fundamental thing that is black but rather blackness comes as study, which is refusing antiblackness, which is desedimentation's impossibility, its future anteriority.*[9] Death is a condition of antiblackness not of blackness because it is a singular event that accounts for life in the form of value; thus *Schwarz-Sein* as the refusal to be born into a separation from the death, from others, is radically flipping the script and living with the death, making death alive, becoming ancestors singing in concert. This is where the idea for the title *Alongside a Chorus of Voices* comes from.

Moving alongside this chorus, to sing in concert with those who cannot sing, who are denied having a voice or those who remain unthought, is to sit with and in blackness. This is a refusal, after Fred Moten and Stefano Harney's thinking of the term in *The Undercommons*, as a sort of doubling down on that which is refused to Black lives (white gaze [times] II?). This refusal is then also, inspired from Gayatri Chakravorty Spivak's articulations of it, an acknowledgment of every individual's relational positionality.[10] That is to say that, as Spivak points out through her example, she has to acknowledge that she is in the context in which she is making the example, not poor and from the upper caste.[11] Thus, refusal brings in the need for the conditions, the positionings or perspectives, from which refusal takes place. But this recognition of the conditions, of contexts, has to be further uprooted. Since a context's own self-containedness and its purely relational existence both reify an omnipotency of itself. The question of refusal thus at once reminds of the need for an elaboration of the condition while at the same time has to allow for a refusal of the world made up of conditions. It is not possible to leave the (antiblack) world behind by simply forgetting (or remembering) the conditions of its existence. Rather, the refusal has to move into a space of refusing the very conditions for refusal. Thus, those who can acknowledge positions must refuse those positions, not so as to leave them behind but so as to know what those conditions inhibit: that which fugitively escapes those conditions.

Such refusal is thus a double-sightedness, hearing through someone else's ears, a breakdown of seeing as one to hearing someone else's sounds inside of yours. This doubling down can be heard under the heading of double consciousness, can be heard as the doubling of the white gaze; that is, a Black person seeing themselves through the eyes of the antiblack world, hearing oneself as other. After quoting Frantz Fanon's own statement of double consciousness—"The black man possesses two dimensions: one with his fellow Blacks, the other with the Whites"[12]—Moten and Harney poetically elaborate:

> Can this being together in homelessness, this interplay of the refusal of what has been refused, this undercommon appositionality, be a place from which emerges neither self-consciousness nor knowledge of the other but an improvisation that proceeds from somewhere on the other side of an unasked question? Not simply to be among his own; but to be among his own in dispossession, to be among the ones who cannot own, the ones who have nothing and who, in having nothing, have everything. This is the sound of an unasked question. A choir versus acquisition, chant and moan and *Sprechgesang*, babel and babble and gobbledygook, relaxin' by a brook or creek in Camarillo, singing to it, singing of it, singing with it, for the bird of the crooked beak, the generative hook of *le petit negre*, the little nigger's comic spear, the cosmic crook of language, the burnin' and lootin' of pidgin, Bird's talk, Bob's talk, bard talk, bar talk, baby talk, B talk, preparing the minds of the little negro steelworkers for meditation.[13]

But refusal can also be heard as affirmation:[14] *affirmation of the refused state that Black Swissness marks—we refuse our refusal to Black Swissness in our affirmation of this refused position, of refusedness. In other words, taking on Black Swissness marks a refusal of Swissness as citizenship because I cannot* be properly present in/as it. *That means what is refused is not the particular case but the general because it is in the exception's unexceptionality, in the way in which the exception is before that which excepts it, that the rule is refused.*[15]

Moten engages the question of refusal and states in another piece titled "Notes on Passage." It is here that the question of home and state is unmasked as a question of Black kinship:

> She is the commodity, the impossible domestic, the interdicted/contradictive mother. Dangerously embedded in the home from which she is excluded, she is more and less than one. The question of where and when she enters—where entrance is reduced to some necessarily tepid mixture of naturalization and coronation, which is an already failed solution that is ever more emphatically diluted in its abstract and infinite replication—is always shaded by the option to refuse what has been refused, by the preferential option not for a place but rather for radical displacement, not for the same but for its change. Blackness is given in the refusal of the refugee.[16]

I want to read this passage after our elaborations and think through the ways in which blackness is already in the state as statelessness, even when not being displaced—their hereness is already displacement. Rather than following displacement as being put into the absent and out of states, I want to think about what happens to Swissness itself in this displacement turned not-exception. This question is important in Switzerland because antiblackness also practices other kinds of displacements: through literal displacements of foreigners or refugees, but also through a kind of endlessly deferred displacement as those who are here and have not only no way into citizenship but also no way to leave.

"Displacement and migration are structurally created as well as maintained,"[17] as Canadian activist and writer Harsha Walia argues. It is not only a result of states' power but how they claim their power, as I have argued here. Furthermore, displacement is not only about being on the move but also about "immobility, preventing both the freedom to stay and the freedom to move."[18] This means that displacement is about the management of noncitizens. Thus, it is the claim of the right to being here that happens through the right to manage those who should not be here that instantiates a proper belonging to a land as a nation. Rather than being outside of the eyes of the state, the state makes itself on the basis of the stateless, where states mark themselves as authority. This control can make some permanently stateless and with only one way out: citizenship.

On the other hand, within citizenship there are those who can never fully belong because they are not *properly* Swiss. This question of blackness is thus, as Moten elaborates, one of the immigrant and the refugee, because they are marked with black until they belong to a state. At the same time, Black people, as blackness in life, always fall short somewhere, someplace, sometime, from being antiblack enough to be proper citizens. But it is in this recognition of Black citizens, which is where citizenship as proper breaks down and thus becomes open to other meanings, that some kind of globality, a world in poetic relation (rather than a series of brutal global states in friendship and enemy relations), comes to the fore, where citizenship as belonging begins to be radically rewritten and borders of nations must change what they are. Here, another modality of citizenship, of home, comes to the forefront. After all, something does happen with each black notebook passed down by ancestors committed to blackness. While living is not a given for Black lives within nations, it is Black lives (and those relegated to the outside of properness) living, writing, musicking, identifying, claiming, and petitioning for and against citizenship that rewrite the world. In this sense I want to add to Moten's "Notes on Passage" and enlarge the notebook of possibility coming out from the black radical tradition that bends this scene to worlds unknown and unthought. That is not about a kind of making of progress, a fixing of a problem, nor about closing knowledge through amassing it. Rather, it is about how study is opened, how other narratives can be written, how learning to make music is a practice of getting to know us that has no end.

We're not writing blackness and Black life into an archive, into a world, into citizenship, into a notion of states, but rather states of existence are transmutated by our black study projects. Blackened states of life, *Schwarz-Sein*, cannot be called by this world or claims to sovereignty except as a continual lack. Enter the door of no return, there where states as stateless are unfounded, where time in multidimensionality is articulated, not as linear nor as nonlinear but as these points of observation where linear time is with those diffracted linearities that are not.[19] Once entered, when you're behind the door of no return, in this blackness, lies the possibility of an alter destiny. Blackness radically shifts that which is thought in motion of oppositions or distinctions, those infinities that meet and make a moment in space. What if we think the world with Black lives? What happens to this world when our decisions flow as a concern for those who are unthought? What if we aren't about me but about others (and I mean that like I'n'I). Thinking with and fighting for Black lives means unraveling the tools to save the planet from its destruction through exploitation; it means making decisions to fight for those whose voices aren't heard in a space; it means asking difficult questions and to do so as a way to celebrate lives in excess of what we might know or think life to be. This is why we need these notes, these musical inspirations toward unheard sounds. Reading after, with, and before notes and other holds (of text, of music) we hear of togetherness in and with Blackness.

Postface
Endless Endlessness

What I aim to elaborate under the headings "Black" and "blackness" in this book exceeds any knowable categorizations within this world. In one way, we can hear this as the question of Black lives, of Black Swiss lives. This is because they do not exist; this is because their existence is a refiguring and remembering of how "Black" has denoted, since the very beginning of the circumscription of this term by antiblackness, an excess of relations—that is, it has always meant more than some kind of delimited kinship, set of traits, or any other antiblack constriction. In fact, I might put it this way: antiblackness closed not-blackness out from *livity* by limiting what life may be and how it may change or be in relation. This delimitation happens in the control of that which embodies, represents, and symbolizes change, relations, lives meeting in excess of the authority it claims.

While I indeed write this here once more as a clarification for those slippages containing Black Swiss in a certain box, as a kind of category of life to be separate from others, I am raising this here again more so because of the conclusion we must draw in relation to antiblackness and blackness. That is, if this is a structuring paradigm, a grammar, an all-encompassing set of rules, if you will, then antiblackness makes itself so by claiming to be above lives. We might say that the antiblack world claims authority over lives; we might say that it can decide on who shall live and who must die; we might say that it defines life and aliveness, and so forth.

Let me explain this with a revisitation of a question that has been with us throughout this book. Importantly, I want to state that this question is here not to be seen as fundamental to the book but rather as a kind of appearance of fundamentalness that it claims. By this I mean that its appearance as necessity is illusory. And by illusory, I mean it is its own ruse, so as to claim its own necessity and authority. It appears as a kind of already closed question: blackness is bespeakable only because of antiblackness. I will now unpack this on the order of two steps: first, the reason why this is in fact a question, and second, why it is in fact wrong—not as the opposite of correct but as the breakdown of the very relations it claims in its form as a question that can be answered.

To do so, I have to revisit and continue, both at the same time, with Charles Uzor and his works for George Floyd. Over the course of writing this book, things continued to change in the world, and in November 2023, Uzor's works for Floyd were curated and presented at the Lucerne Festival Forward. Specifically, they programmed Uzor's *8'46" George Floyd in Memoriam* and commissioned a new work titled *Katharsis Kalkühl: George Floyd in Memoriam*. But *Bodycam Exhibit 3: George Floyd in Memoriam* was not performed. According to a newspaper article: "For some young LFCO members, this was too much emotionally. There was an open discussion about it; in the end, it was decided that the second piece from Uzor's Floyd trilogy would not be played in Lucerne."[1] For Uzor, this was understandable, as the article mentions.[2] While I am not able to comment on the workings at the festival, the performance, or the circumstances, since the event took place at the very end of this book's writing process, this brief article requires us to engage in a discourse that appears as the question: Is blackness bespeakable only because of antiblack violence?

Specifically, in the field of Black studies in the United States, there appears currently a split between two headings: "optimism" versus "pessimism." I can't join this discourse of dichotomy. To me, the case of Black Swiss radically includes all the views around blackness, as I have shown, and asks us to think beyond all that has ever been sayable/thinkable. The study of Black Switzerland shifts the very questions involved.

The newspaper article covering the 2023 Lucerne Festival Forward ends with a series of questions: "Is it permissible to impose such violent original documents on the public and the performers? Are the relatives or even the victim himself instrumentalized by such a representation? How far should art be allowed to go?"³ What is revealed is that the avoidance of the telling of this event works too easily into and alongside a discourse of color-blindness and a claimed absence of antiblackness in Switzerland and Europe. As I elaborated in chapter 1, following Steve Martinot and Jared Sexton, it is the turning into an exception of the killing of Black lives that not only erases the very reason for such brutalities—that is, antiblackness—but also instantiates it.⁴ That is, what comes to the fore is that the not telling about the killing of Black lives happens by way of marking not only the police as exception to an otherwise good world but the whole event. To put it simply: what is forgotten, as revealed by these questions, is that Floyd was killed because he was Black. Furthermore, it also implies that this killing contains antiblackness—it appears only in the event of murder. And lastly, the event can be placed outside of here, as something that does not happen in Switzerland and is only the fault of some "rogue cops."⁵

But the article poignantly concludes with Uzor's statement that "we still hear about similar cases of police violence against black people today. This does not stop."⁶ Still, we must add that it also happens in Europe and also in Switzerland. And yet this still does not address that antiblackness is not reducible to the event of murder, for that would still mark these moments as exceptions and forget that they are killed because they are Black. Saidiya Hartman's work extends our understanding of this scene to encompass more than this event, that the violence of antiblack brutalities is found everywhere, even in "the mundane and quotidian."⁷

Importantly, what this scene proves is that the avoidance of voicing the killing of Black lives is a necessity for antiblackness. Antiblackness requires its own articulation to be absolutely uncontestable; thus it must erase its own formation. Such erasure is only possible if the violence is containable in a self-contained event, in a separated instance. This is antiblackness's claim to control the movement of narration, the movement of time, toward its own becoming and absolute presence. That is also why Hartman's not-telling is a kind of telling, a diffusion of the telling.⁸ Furthermore, this means that Hartman is telling us of a scene of antiblack brutality and uncovering not only its naturalization and normalization in the form of the portrayal of an actual scene of murder but also how it is so naturalized and normalized that this brutality is everyday. Thus, not only does the circulation of these kinds of scenes in their consumption become naturalized and might fail to show the violence of antiblackness, but rather this violence is *always* naturalized and normalized by antiblackness—even in the event of murder itself.⁹ And, antiblackness must continually make this scene in various ways, beyond the act of murder, to be able to naturalize and feed off of it. To normalize itself, antiblackness erases its own instantiation by claiming it is natural through the figure of the other—or, more precisely, blackness is a reservoir of naturality.

In Switzerland, a raceless space, antiblackness continually erases the event of actual brutality by reducing it not only to the scene of police violence or the killing of Black lives, but also by displacing it somewhere else. All forms of antiblackness are placed somewhere far away from its own domain. This is what Switzerland teaches us as a general principle of antiblackness. In the United States, this might not appear to be the case initially, but it is in hearing Hartman in this vein that we can uncover this pervasiveness of the naturalization of antiblack violence as always already necessarily claimable by antiblackness at different borders to its sovereign space—including all those things that fall into the category of the everyday.

Here our closed question opens: Is blackness only bespeakable in the face of antiblackness? This question was hidden by antiblackness's always already proclaimed answer to it. In fact, the question itself already supposes an answer, as all answers already are a kind of yes (and even if the answer is no, its articulation takes the form of having to answer to antiblackness, thus still a kind of yes, some might say). But this is why this scene moves us even further from antiblackness's self-proclaimed authority. What does it mean when there is no way to hear any of these sounds outside of the antiblackness it bespeaks? While we could hear then how there is always already blackness in excess of antiblackness, because blackness always must be breaking antiblackness, this still requires us to point out one important consequence: if blackness is in excess of antiblackness, then it is not blackness that continually resists but rather antiblackness is continually attempting to hold blackness, to arrest its movement, to stop its musicking.¹⁰ In other words, the scene of violence is always already normalized and appears only as violence in and with antiblackness's aim to regain control over its own instantiation. Black lives are killed and become absolute localities of blackness, because antiblackness must claim control over blackness. But this implies even a further step: antiblackness is trying to make itself in moments of violence, where it can exercise control. This is why it must perform this violence everywhere to uphold itself.

But *antiblackness*, as a term, bespeaks exactly this claim to exercising control, this claim to all-encompassing authority, this claim that *is* brutalities against Black lives and blackness. For as the word reveals, antiblackness is the attempt to hold blackness off, to hold lives in containers, to stop lives from meeting, to stop people from moving in and out, to arrest the music, to authorize meaning, to control the flow of time. Blackness thus becomes a term that opens unthought, unspoken questions in its refusal to the claim over the definition of what an instance may be.

In Uzor's presence, in the article and the festival and the discussions about the performance of the work, is engaged always already the event that is aimed to be not reproduced. This means that it is there, as an absent, structuring principle. But this also means that regardless of whether there is a portrayal of this violence or not, the critique of what the event presents and that which exceeds the critique is always already there.[11] This always already thereness is not reducible to its emergence by a pressure, or violence imposed by antiblackness, nor to a staticness of a kind of unmoving thing. Rather, it exceeds both its relativity and its absoluteness as the very breakdown of a narrative of causality, of relations.

Blackness thus must exceed antiblackness as a particular articulation in our historical epoch, as existence as one name. As I mentioned, we must keep up our work, our wake work, for Mike Ben Peter: the judicial systems' verdict that his death was not provable as the police's liability reminds us how we must keep insisting that there is antiblackness here—how the erasure of Black continues.[12] But in our continual protest, in our continual study, in our continual musicking and bringing to discussion the case for Black lives, a case that seems to remain inaudible to an antiblack world, Blackness bespeaks more than antiblackness could ever mean or do. As a banner reads in an article documenting the continued fight for justice by Peter's family: "Mike Ben Peter. Rest in Power. We Will Never Forget You."[13] Blackness, as something like a practice, bespeaks an endless endlessness of thinking and sounding: it bespeaks the way in which endings exceed all beginnings and endings. Beyond this book's last pages, we *still* keep on—Blackness announces our endless endlessness.[14]

Notes

INTRODUCTION. BLACK SWISS

1. Referencing Davis, *In a Silent Way*.
2. See, for example, Adu-Sanyah's description of her *Inheritance—Poems of Non-Belonging* series, of which the piece *White Gaze II* is a part. As she writes: "Inheritance—Poems of Non-Belonging is a growing narrative on the incompatible nature of my biracial identity, provoked by the conflicting desires of belonging for once, and yet wanting to be able to define the Self independently from external judgment and categorization, while remaining the other—regardless if I'm confronted with my presence in the white culture I grew up in, or with a temporary appearance in a Black culture." Adu-Sanyah, "Current Work."

    It is exemplified in the question "Where are you from?," a question that implicitly makes the proper here as not-Black, nor any other people of color, as white. See Zanni, "Die Frage 'Woher kommst du?' ist rassistisch."
3. Campt points out the similar situation in Germany in the twentieth century. Campt, "Afro-German Cultural Identity and the Politics of Positionality," 113.
4. See Wright, "Others-from-Within from Without," 298.
5. The earliest documented organization I am aware of at this time that centered Black lives in Switzerland was Women of Black Heritage (WBH), founded in the late '80s. Burke et al., *I Will Be Different Every Time*, 58.
6. For a good reference, see the website Black Central Europe, "Home."
7. "To All Art Spaces in Switzerland," Black Artists and Cultural Workers in Switzerland.
8. "To All Art Spaces in Switzerland," Black Artists and Cultural Workers in Switzerland. Also, as Stevens notes, there are many more cases. Stevens, "Many Black People Get Killed by Police in Quiet Switzerland Too."
9. "Switzerland Must Urgently Confront Anti-Black Racism," United Nations.
10. "Expertengruppe zu Rassismus," SRF Schweizer Radio und Fernsehen.
11. Martinot and Sexton, "The Avant-Garde of White Supremacy."
12. See, for example, Jung and Vargas, *Antiblackness*, 8.
13. Kay, "Berlin-Based Producer Bonaventure on Trauma,."

14. Bonaventure, "Supremacy."

15. I am here referencing Du Bois's practice. See Wright, *Becoming Black*, 74.

16. Campt elaborates on how the term *Afro-Germanness* from its inception was not delimitable into a distinct identity category but rather articulated as a "plurality" of identities, or heritages and experiences—from race, to nationality, and gender. Campt, "Afro-German Cultural Identity and the Politics of Positionality," 111.

17. Black Europeans are keenly aware of this modality of antiblackness. See, for example, Batumike, *Noirs de Suisse*, 127; Campt, "Afro-German Cultural Identity and the Politics of Positionality," 123–24; Singletary, "Everyday Matters," 148.

18. Du Bois's choice for the term "Negro" as a modification of an antiblack definition exemplifies such. While Du Bois does mention it to describe a people, his elaboration initially and the following questions opened by his elaboration exceed a notion of a properly delimited group of people. Du Bois, *Writings*, 1220–22.

19. Batumike discusses racism and also briefly the shared experience of Black lives across boundaries of national origin/belonging in Switzerland in *Présence africaine en Suisse*, 40–44.

20. Batumike, *Présence africaine en Suisse*, back cover; L'Harmattan, "Cikuru Batumike."

21. Scholars of Afro-Germany have pointed out the hegemony of US discourses within Black studies. Campt, *Other Germans*, 176–78.

22. Burke et al., *I Will Be Different Every Time*.

23. I'm referencing here the mystical ruminations around Black lives in the Atlantic Ocean, because many Black lives were lost in the ocean during the slave trade. See, for example, Gumbs, *Undrowned*.

24. I'm leaning here on Moten's insistence on staying within the gap of translating Fanon's fifth chapter of *Black Skin, White Masks*, in "The Case of Blackness," 179–80.

25. As Gordon elaborates:

> In effect, that to see what G-d (in such a presumption) saw was to see that which should not be seen. There is a form of illicit seeing, then, at the very beginnings of seeing black, which makes a designation of seeing in black, theorizing, that is, in black, more than oxymoronic. It has the mythopoetics of sin.
> Although the subsequent unfolding of theory claimed other sites of legitimacy, where G-d fell to systems of thought demanding accounts of nature without an overarching teleology—instead elevating what could be thought through inescapable or insurmountable resources of understanding, as Kant subsequently argued—the symbolic baggage of prior ages managed to reassert themselves at subterranean levels of grammar. (Gordon, "Theory in Black," 197)

26. Like the way composer Muhal Richard Abrams would tune his ensembles:

> At the beginning of the second rehearsal, Abrams suggested that the brass and woodwinds "tune up." In place of the usual single tuning note, however, Abrams looked down at some paper on his music stand and started assigning pitches to players. The musicians, who probably expected to tune in a conventional fashion, were suddenly scrambling to make sense of Abrams's directions. It gradually emerged that Abrams was constructing four thirteen-note chords, which

he described as alternating between having "tension" and "no tension," and that these chords and their progression constituted his "tune up": "let's see what kind of sound we're getting as a unit," he said immediately before the band played through the chords. (See Hannaford, "One Line, Many Views," 1–2.)

27. And thus the book performs a radical refusal of a certain kind of narrative that is antiblack in its delimitation of time and events. See chapter 1.

28. Like it used to be, and still is, depending on how we define community, the case in Germany: "Black Germans are not a community, and also not an outside." See Campt, *Other Germans*, 103

29. Jazz is grouped with all musical styles that are not classical (or *traditional*) in the musical pedagogical system in Switzerland. For an elaboration of the way that the production of whiteness and classical music intersect, see Thurman, "Performing Lieder, Hearing Race."

30. See chapter 5 in Fanon, *Black Skin, White Masks*.

31. Leaning on Moten's elaboration of falling, black squares, and Blackness in "The Case of Blackness."

32. Uppercase *Black* denotes all things to do with Black lives, while *blackness* denotes a concept. I have elsewhere attempted to reopen the question of this continual movement. Cox and Jean-François, "Aesthetics of (Black) Breathing," 106, 108, 110.

33. That is, if blackness is absolutely knowable and encircleable by an antiblack world. We might hear this as a matter of tokenism, where race and otherness are contained within one, or a few, persons of color and, at the same time, whiteness continues to function as the norm. See El-Tayeb, *European Others*, 6.

34. It must be so, for its reduction to mere race is the way antiblackness protects itself, or distances itself, from blackness.

35. We might think of this as how blackness asks us to refigure the human or, alternatively also, as how blackness asks us to think in excess of the human. I am leaning here also on Moten's observation that "blackness cuts the distinction between essence and instance." Moten, *Stolen Life*, 15–16, 241.

36. I hear one of the authors of *I Will Be Different Every Time* transform the notion of the individual when she resists being reduced, as a Black person, to being the voice/image of all Black people (from a question of being an individual on the basis of opposing others to being an individual as the others' refusal of being reduced to one thing). Burke et al., *I Will Be Different Every Time*, 23.

37. Sun Ra, "Death Speaks to the Negro," 0:40–1:20.

38. In poetic relations, to be exact, which requires us to think of relations in excess of relations, as some kind of opaque poetics. After Glissant, *Poetics of Relation*.

39. Moten uncovers how blackness is a generativity that bursts the seams of language and law or structure. Moten, "Jurisgenerative Grammar (for Alto)."

40. Beyoncé, "Me, Myself and I."

CHAPTER 1. INTERSTITIAL LISTENINGS I

1. Judy points out that methodology is entangled in domains of knowledge and how Blackness requires a refiguring of them. Judy, *(Dis)forming the American Canon*, 3–5.

2. AFP/The Local, "Protests Force Switzerland to Grapple with Its Own 'George Floyd' Case"; Fernandez, "Affaire Mike Ben Peter."

3. Swiss Confederation, "Federal Act on Foreign Nationals and Integration," art. 89.

4. Dos Santos Pinto et al., *Un/Doing Race*, 12.

5. Cox, "Stories of the Mothership."

6. "Logothetis 100.3," Wien Modern.

7. A third one in the series, *Katharsis Kalkül: George Floyd in Memoriam*, premiered during the completion of the revisions of this book on November 19, 2023, at the Lucerne Festival Forward. "Ensemble of the Lucerne Festival Contemporary Orchestra (LFCO)," Lucerne Festival.

8. Uzor, *Bodycam Exhibit 3*.

9. See Cox and Jean-François, "Aesthetics of (Black) Breathing," 101–2.

10. Hartman, "Venus in Two Acts," 5.

11. Hartman, "Venus in Two Acts," 11.

12. Willis et al., "New Footage Shows Delayed Medical Response to George Floyd."

13. Leaning here on Spivak, "Can the Subaltern Speak?," 69–74.

14. Martinot and Sexton, "The Avant-Garde of White Supremacy," 178–79.

15. Stadler, "Never Heard Such a Thing," 87.

16. Stadler, "Never Heard Such a Thing," 89, 94.

17. Stadler, "Never Heard Such a Thing," 98.

18. Stadler, "Never Heard Such a Thing," 99.

19. For further elaboration, where the authors discuss this in relation to Morrison's seminal work on Blacksound, see Cox and Jean-Francois, "Aesthetics of (Black) Breathing," 106–10; Morrison, "Sound in the Construction of Race"; Morrison, "Race, Blacksound, and the (Re)Making of Musicological Discourse."

20. Sutherland, "Making a Killing."

21. Ablinger, "Quadraturen."

22. Barrett, "Between Noise and Language," 159.

23. Wilderson, *Red, White & Black*, 26–27.

24. The pitches don't reduce to so few as at the beginning, but there are many more instruments.

25. In the recording from the concert at Wien Modern by oenm, at 3:45–5:20.

CHAPTER 2. BLACKNESS AND BLACK LIVES IN SWITZERLAND

1. An example is Université Populaire Africaine En Suisse (UAPF), cofounded in 2008 by Mutombo Kanyana, who also ran the magazine *Regards Africains* from 1986 to 2005. "Dr. Mutombo Kanyana," Schauspielhaus Zürich; Université Populaire Africaine En Suisse, "Home"; Collectif Afro-Swiss (CAS), "Cas quoi . . . ?"

2. I elaborate this in relation to Switzerland here, but for a similar critique regarding Germany, see El-Tayeb, *Schwarze Deutsche*.

3. See, for a discussion, dos Santos Pinto, "Spuren," 145–46.

4. Nadel makes a similar point: "The South rather accommodated the guilt of failure by merely segregating the sign of that failure—the ostensibly free slave—making it an otherness, removing it into silence and invisibility in much the way that, in the

preceding century, Europe segregated the mad. But this shift created a new need. Physically free, the black through invisibility determined the limits of the institution. The endurance of the institution, therefore, required new ways of maintaining the invisibility of the black; these were found by moving blacks from physical bondage into silence." Nadel, *Invisible Criticism*, 11.

5. Batumike, *Être Noir africain en Suisse*, 80.

6. See, for further examples, the Alliance against Racial Profiling, "The Alliance"; Schwarze Schweiz, "Archive."

7. Weheliye, *Habeas Viscus*, 17–19.

8. See Lewis, "A Small Act of Curation."

9. Forschungskollektiv Rassismus vor Gericht, "Racial Profiling vor Gericht," 1.

10. Swiss Confederation, "Federal Act on Foreign Nationals and Integration," art. 89.

11. See Naguib, "Mit Recht gegen Rassismus im Recht."

12. Kreis, "Bundesgericht entlastet SVP-Politiker."

13. Boulila, "Race and Racial Denial in Switzerland," 1401.

14. Forschungskollektiv Rassismus vor Gericht, "Racial Profiling vor Gericht," 1.

15. Swiss Confederation, "Federal Constitution of the Swiss Confederation," art. 8, abs. 2.

16. Young, "Rassismus for Gericht," 2.

17. Murakawa, *The First Civil Right*.

18. Boulila, "Race and Racial Denial in Switzerland," 1407.

19. Forschungskollektiv Rassismus vor Gericht, "Racial Profiling vor Gericht," 11.

20. Forschungskollektiv Rassismus vor Gericht, "Racial Profiling vor Gericht," 15.

21. Bonilla-Silva, *Racism without Racists*, 2.

22. Leaning on Bonilla-Silva's four frames of color-blindness. Bonilla-Silva, *Racism without Racists*, 26–28.

23. What Bonilla-Silva terms "abstract liberalism": "Using ideas associated with political liberalism (e.g., 'equal opportunity,' the idea that force should not be used to achieve social policy) and economic liberalism (e.g., choice, individualism) in an abstract manner to explain racial matters." Bonilla-Silva, *Racism without Racists*, 28.

24. Bonilla-Silva's frame "naturalization": "explaining racism's effects away through claiming that they're natural." Bonilla-Silva, *Racism without Racists*, 28.

25. Somewhat related to the frame of minimalizing racism: claiming that racism is not relevant to minorities anymore, as racism has been overcome. What is different is that it is not even necessary to claim that racism isn't relevant to "minorities," but through its more pronounced erasure, the question of its relevance is unaskable and always already answered in a general, even when it is a matter of relevance for an individual that is made to be the loci of race and racism. Bonilla-Silva, *Racism without Racists*, 29; Forschungskollektivs Rassismus vor Gericht, "Racial Profiling vor Gericht," 4.

26. As Michel mentions, the idea of racelessness facilitates racial profiling since this subverts any possibility of critiquing racial profiling. Michel, "Racial Profiling und die Tabuisierung von 'Rasse,'" 112.

27. Forschungskollektiv Rassismus vor Gericht, "Racial Profiling vor Gericht," 15.

28. Forschungskollektiv Rassismus vor Gericht, "Racial Profiling vor Gericht," 15–16.

29. Wa Baile et al., *Racial Profiling*.

30. For a discussion of the group, see Bruggmann et al., "Draussen—drinnen—dazwischen."

31. Biographic information from Fröhlicher-Stines, "The Effects of Racism on Group and Individual Identity"; Stadt Zürich Präsidialdepartement, "Stadttaler an Carmel Fröhlicher-Stines und an Zeedah Meierhofer-Mangeli."

32. Fröhlicher-Stines and Mennel, *Schwarze Menschen in der Schweiz*, 7.

33. See Bernasconi, "Who Invented the Concept of Race?"

34. Weheliye, *Habeas Viscus*, 50.

35. See hooks, "Eating the Other."

36. Browne, *Dark Matters*, 136.

37. See, for example, Dancy and Saucier, "AI and Blackness."

38. For an elaborative survey on AI bias, see Garcia, "Racist in the Machine." For an elaboration of how the combating of AI bias requires investigating the structures (institutions, communities, etc.) behind their creation, see Dancy and Saucier, "AI and Blackness."

39. For an elaboration of a use of data in the form of population demographics in the related context of France, and how choices in what and how measurements of populations are entangled in questions of race, see Simon, "The Choice of Ignorance," 65–66.

40. Meyer, "Prävention gegen die Diskriminierung bei Personenkontrollen durch das Grenzwachkorps."

41. Meyer, "Prävention gegen die Diskriminierung bei Personenkontrollen durch das Grenzwachkorps."

42. Meyer, "Prävention gegen die Diskriminierung bei Personenkontrollen durch das Grenzwachkorps."

43. Meyer, "Prävention gegen die Diskriminierung bei Personenkontrollen durch das Grenzwachkorps."

44. Michel, "Sheepology," 410–11.

45. Michel, "Sheepology," 411.

46. Michel, "Sheepology," 417.

47. Michel, "Sheepology," 418.

48. Translated and cited by Michel in "Sheepology," 418; "Le Parti Radical répond aux affiches polémiques de l'UDC montrant un mouton noir 2007," RTS Radio Télévision Suisse.

49. Michel, "Sheepology," 419.

50. Michel, "Sheepology," 417.

51. Translated and cited by Michel, "Sheepology," 419; Conseil Municipal, "Mémorial," 955.

52. Michel, "Sheepology," 419.

53. Yancy elaborates, after Sara Ahmed, that "whiteness is invisible to those who inhabit it." Furthermore, Yancy points out how such is a limitation on knowledge due to a belief to be able to always have absolutely transparent access to everything. Yancy, *Look, a White!*, 7, 171; Ahmed, "Declarations of Whiteness."

54. Michel, "Sheepology," 417.

55. Swiss Confederation, "Federal Act on Swiss Citizenship." For a critique of *jus solis* in neighboring Italy in relation to race, see Muvumbi, "Black Lives Matter in Italy."

56. Swiss Confederation, "Federal Act on Swiss Citizenship." I also discuss this briefly in relation to Uzor's work in Cox, "Stories of the Mothership."

57. Swiss Confederation, "Swiss Civil Code," art. 97a.

58. Swiss Confederation, "Swiss Civil Code," art. 97a.

59. Lavanchy, *How Does "Race" Matter in Switzerland?*, 10.

60. Lavanchy, *How Does "Race" Matter in Switzerland?*, 11.

61. Lavanchy, *How Does "Race" Matter in Switzerland?*, 14.

62. Lavanchy, *How Does "Race" Matter in Switzerland?*, 14.

63. Sexton, *Amalgamation Schemes*, 218.

64. Sexton, *Amalgamation Schemes*, 218.

65. Sexton mentions this in a similar manner in *Amalgamation Schemes*, 221.

66. Sexton, *Amalgamation Schemes*, 218.

67. Cretton, "Performing Whiteness," 854–55; Lavanchy, *How Does "Race" Matter in Switzerland?*, 11, 15–16, 19.

68. Cretton, "Performing Whiteness," 855.

69. Cretton, "Performing Whiteness," 855.

70. Cretton, "Performing Whiteness," 856.

71. Cretton, "Performing Whiteness," 855.

72. Cretton, "Performing Whiteness," 855.

73. See also Mbembé, *Critique of Black Reason*, 49.

74. Elaborated in Chandler. "Beyond This Narrow Now."

75. Stovall, "The Color Line Behind the Lines," 767.

76. Stovall, "The Color Line Behind the Lines," 767.

77. Cretton, "Performing Whiteness," 843.

78. See Purtschert and Fischer-Tiné, *Colonial Switzerland*; Purtschert et al., "Switzerland and 'Colonialism without Colonies'"; Lüthi et al., "Colonialism without Colonies"; Purtschert, "The Return of the Native"; Cooperaxion, "Datenbank der im Sklavenhandel involvierten Schweizer."

79. Cretton, "Performing Whiteness," 843.

80. Cretton, "Performing Whiteness," 853.

81. Cretton, "Performing Whiteness," 855.

82. Gondola, "'But I Ain't African, I'm American!,'" 209.

83. Batumike also problematizes integration and inequality, particularly how it implies that those who already are Swiss never (have to) change. Batumike, *Noirs de Suisse*, 58.

84. Swiss Confederation, "Federal Act on Swiss Citizenship," art. 12.

85. Batumike, *Être Noir africain en Suisse*, 52.

86. Of course Fanon makes a similar point in *Black Skin, White Masks*. In the contemporary noncolonial context, scholars such as Coulthard elaborate on this in relation to Indigeneity in Canada. Coulthard, *Red Skin, White Masks*.

87. Ayim, *Grenzenlos und unverschämt*, 92.

88. Abraham, "The Life and Times of Anton Wilhelm Amo"; Oguntoye et al., *Showing Our Colors*, 3–4.

89. Oguntoye et al., *Showing Our Colors*, 3–4.

90. Oguntoye et al., *Showing Our Colors*, 41–49.

91. Oguntoye et al., *Showing Our Colors*, 59–53.

92. Oguntoye et al., *Showing Our Colors*, 79–94.

93. Walcott, *The Long Emancipation*, 55.

94. Walcott, *The Long Emancipation*, 55.

95. Hartman, *Lose Your Mother*, 148.

96. Marriott makes a similar point in relation to the psychoanalytic theory of Lacan in "Ontology and *Lalangue*," 229–31.

CHAPTER 3. INTERSTITIAL LISTENINGS II

1. Jolo, "Perspectives and Blurred Colors."

2. Of course, groove in fact should always be this kind of excess of the grid. We might also say rhythm should be, too, if we follow drummer Milford Graves. See Cox and Jean-François, "Aesthetics of (Black) Breathing," 104, 114.

3. See Glissant, *Poetics of Relation*, 189–91.

4. Jolo, "Als Kind wünschte ich mir eine Zeit lang, weiss zu werden," 3–4. Translation by Jessie Cox.

5. I use this term after Moten, who takes it from Cedric Robinson. In my view it bespeaks a kind of underground theory, as both resistance to antiblackness but also as the very generativity of thought and theorizing that requires engagement with blackness. Moten, "Not in Between."

6. Barad, *Meeting the Universe Halfway*, 179.

7. Barad, *Meeting the Universe Halfway*, 302.

8. Jolo, "RED, 2018."

9. Kukuruz Quartet and Julius Eastman, *Piano Interpretations*.

10. Jolo, "RED, 2018."

11. I am here leaning on both my own earlier use of this term as well as Moten's. Moten, "Not in Between"; Cox, "Cecil Taylor's Posthumanistic Musical Score."

12. Jolo, "RED, 2018."

13. Glissant, *Poetics of Relation*, 94.

14. Leaning on Moten, who thematizes a similar case with "unasking." He posits this need to unask when citing the question that Wilderson poses: "Would nothing ever be with nothing again?" Where nothing equates with Black life. But for me this unasking is already the question. This is, on the one hand, on the register of Black lives, how our not knowing Black lives operates in the interstitial space that is the study of Black lives, but also as an extension in relation to nothing in general. More on that in chapter 10. Moten, "Blackness and Nothingness," 756. Quoting Wilderson, *Incognegro*, 265.

15. See an elaboration of this in Chandler, "Beyond This Narrow Now," 224.
16. Du Bois, "My Evolving Program for Negro Freedom," 57.
17. Du Bois, "My Evolving Program for Negro Freedom," 56.
18. See Broad, "The Social Register."
19. Sutton, "Direct Me NYC 1786."
20. See, for example, Goldstein, "City Directories as Sources of Migration Data."
21. Hill, *The Book of Negroes*; Krampe, *The Past Is Present*.
22. Du Bois. "Sociology Hesitant." For an elaboration of the history of this essay, see Judy, "Introduction."
23. Du Bois, "Sociology Hesitant," 41.
24. Du Bois, "Sociology Hesitant," 42.
25. From Nativ's song "Sira." Nativ, "Diese Revolution beginnt in unseren Köpfen."
26. Degonda, "Rapper Nativ macht sich gegen Rassismus stark."
27. Degonda, "Rapper Nativ macht sich gegen Rassismus stark."
28. See, for example, in related germanaophone context Kusmierz, "Areas of Uncertainty"; Layne, *White Rebels in Black*; Thurman, *Singing like Germans*.
29. See, for example, Ndiaye, *Scripts of Blackness*.
30. Batumike, *Noirs de Suisse*, 90.
31. Leaning on Morrison's elaboration in the US context. Morrison, "Race, Blacksound, and the (Re)Making of Musicological Discourse."

CHAPTER 4. AFROFUTURIST ARCHEOLOGY

1. See Khan, "Dehumanisation, Animalisation."
2. See, for example, Bundi, "Schwarze Schweizer über Rassismus."
3. Such as who is scared of the black man. In the context of France and children's literature, see Mbembé, *Critique of Black Reason*, 64.
4. "Jovita dos Santos Pinto über Alltagsrassismus," SRF Schweizer Radio und Fernsehen.
5. Fanon, *Black Skin, White Masks*, 85.
6. Or "epidermalizing." See Fanon, *Black Skin, White Masks*, 4.
7. Fanon, *Black Skin, White Masks*, 82–83.
8. Fanon, *Black Skin, White Masks*, 82.
9. Fanon, *Black Skin, White Masks*, 82–83.
10. I borrow the notion of "negative other" from Wynter, "'We Know Where We Are From,'" 6.
11. I read Fanon proposing such and tie it, through my phrasing, to Du Bois's double consciousness. "The black man has two dimensions. One with his fellows, the other with the white man." Fanon, *Black Skin, White Masks*, 8, 114–15.
12. Fanon, *Black Skin, White Masks*, 8.
13. In the novel *Darktown Strutters*, the character Jim Too declares: "'Blackin up is US DOIN WHITE FOLKS DOIN US!'" W. Brown, *Darktown Strutters*, 138. For a discussion of this novel and its thematizing of minstrel blackface performances and the performance of race, as well as how blackface is a performance of whiteness, see Nowatzki, "'Blackin' Up Is Us Doin' White Folks Doin' Us.'"
14. Fanon, *Black Skin, White Masks*, 8.

15. Fanon, *Black Skin, White Masks*, 82–83.

16. Fanon, *Black Skin, White Masks*, 82–83.

17. Judy, "The Unfungible Flow of Liquid Blackness," 31.

18. Cox and Jean-François elaborate alongside/after Farred the way that blackness in its liquidity exceeds states of matter and thus also solidification. Cox and Jean-François, "Aesthetics of (Black) Breathing," 108, 114–15; Farred, "Daseinstufe," 41, 50.

19. Black has no resistance to the gaze as Fanon elaborates. Fanon, *Black Skin, White Masks*, 83.

20. Fanon speaks in particular of "ontological resistance." Fanon, *Black Skin, White Masks*, 82–83. It is my view that Fanon does so as a performative methodology—that is, his writing is to me a kind of performance, and like such enacts a kind of transmutation in the reader.

21. I am thinking here also of hailing in an Althusserian sense. Althusser, "Idéologie et appareils idéologiques d'État."

22. Sylvia Wynter hints at this in her own elaboration of the sociogenic principle and the white gaze in Fanon. Wynter, "Towards the Sociogenic Principle."

23. I am leaning here on Akomfrah, *The Last Angel of History*, and Eshun, "Further Considerations of Afrofuturism."

24. Coney, *Space Is the Place*, 0:00–0:40.

25. I'm referencing what the Arkestra calls "space chords." Szwed, *Space Is the Place*, 214.

26. I am required to use this term, for it is not through identity nor identification that whiteness articulates its identity, its all-encompassing plurality of identities. At the same time, I am aware of a kind of use of this term as a refusal of such orders of worlding where disidentification becomes a disruptive force. Albeit to me, this is a question about the double, about doubling down on disidentification, at least on the order of those who are always already the unidentifiable. See Muñoz, *Disidentifications*.

27. Agamben, *Homo Sacer*, 19.

28. Agamben, *Homo Sacer*, 99.

29. Agamben's definition of sovereignty—that the sovereign is the authority to decide on the state of ecxeption, as well as the friend-enemy relation as basis of the political—are borrowed from Schmitt, *Political Theology*, and Schmitt, *The Concept of the Political*.

30. Weber, "Bare Life and Life in General," 11.

31. Weber, "Bare Life and Life in General," 11.

32. "*The rule applies to the exception in no longer applying, in withdrawing from it.* The state of exception is thus not the chaos that precedes order but rather the situation that results from its suspension. In this sense, the exception is truly, according to its etymological root, *taken outside* (*ex-capere*), and not simply excluded." Agamben, *Homo Sacer*, 17–18.

33. Weber, "Bare Life and Life in General," 15.

34. Agamben, *Homo Sacer*, 90.

35. It is fascinating that Agamben is able to point out the "indeterminateness" of the separation only when thinking through the case of having to join citizenship and leaving something else behind. Agamben, *Homo Sacer*, 90.
36. Agamben, *Homo Sacer*, 111.
37. Weber, "Bare Life and Life in General," 14.
38. Schinkel, "From Zoëpolitics to Biopolitics," 156.
39. As Schinkel elaborates:

> Zoëpolitics is primarily externally directed towards persons outside the state, as becomes visible, for instance, in the reduction to bare life of those detained in Guantanamo Bay and in the administrative detention of "illegal aliens." Biopolitics is a second form of biopower. It is internally directed and aims at the control of populations occupying the state's territory but which are discursively placed outside the domain of hegemony marked as "society." Biopolitics takes as its object the social body, the *bios* that is usually referred to as "society." It involves the sorting of populations according to who is deemed part of "society" and who isn't. (Schinkel, "From Zoëpolitics to Biopolitics," 156)

40. Schinkel, "From Zoëpolitics to Biopolitics," 158.
41. Schinkel, "From Zoëpolitics to Biopolitics," 158.
42. Schinkel, "From Zoëpolitics to Biopolitics," 158.
43. Schinkel, "From Zoëpolitics to Biopolitics," 159.
44. Weheliye, *Habeas Viscus*, 53.
45. Weheliye, *Habeas Viscus*, 54–55.
46. Weheliye, *Habeas Viscus*, 54.
47. Agamben, *Remnants of Auschwitz*, 85.
48. Weheliye, *Habeas Viscus*, 55.
49. Weheliye, *Habeas Viscus*, 55–56.
50. Weheliye, *Habeas Viscus*, 57.
51. Weheliye, *Habeas Viscus*, 71.
52. Weheliye, *Habeas Viscus*, 71.
53. For an elaboration of how Blackness is claimed as absolutely outside, how the very making of the human requires this kind of violence of border control against Blackness, see Wilderson, *Afropessimism*.
54. Oguntoye et al., *Showing Our Colors*, 79–94. See, for a detailed study, Lusane, *Hitler's Black Victims*.
55. Busey and Dowie-Chin, "The Making of Global Black Anti-Citizen/Citizenship," 158.
56. Brackets in citation. Busey and Dowie-Chin, "The Making of Global Black Anti-Citizen/Citizenship," 158.
57. Mbembé, "Necropolitics," 23.
58. Mbembé, "Necropolitics," 17.
59. Mbembé, *Necropolitics*, 39–40.
60. Mbembé, *Necropolitics*, 15–16.
61. Mbembé, *Necropolitics*, 16.
62. Mbembé, *Necropolitics*, 16.

63. Mbembé, *Necropolitics*, 77.

64. Mbembé, *Necropolitics*, 21. Akin to, although not quoted as a reference, Patterson's seminal study *Slavery and Social Death*.

65. Providing a critique of this tendency in the concept of necropolitics, see Sexton, "People-of-Color-Blindness."

66. Mbembé, *Necropolitics*, 91.

67. Mbembé, *Necropolitics*, 91.

68. Mbembé, *Necropolitics*, 91.

69. Wilderson, *Red, White & Black*, 82.

70. Weber, "Bare Life and Life in General," 20–21.

71. Weber, "Bare Life and Life in General," 22.

CHAPTER 5. INTERSTITIAL LISTENINGS III

1. Chénière, *Sonic S.cape*.

2. L'Abri, "Maïté Chénière."

3. Chénière, *Sonic S.cape*.

4. I'm thinking here with and after Glissant's notion of *tout-monde*. Glissant, "From the Whole-World Treatise."

5. Glissant's elaboration of, and in, *Poetics of Relation* similarly begins with the abyss of the slave ship. Glissant, *Poetics of Relation*, 5–9.

6. Du Bois, "Sociology Hesitant."

7. Dos Santos Pinto, "Spuren," 160–62.

8. Dos Santos Pinto, "Spuren," 164–65.

9. Dos Santos Pinto, "Spuren," 165.

10. In this way it is following in the footsteps of what Chandler also uncovers in Du Bois: a radical incursion in the very telling of history as narrative that defines our world, lives, and thought. Chandler, *"Beyond This Narrow Now,"* 133.

11. Dos Santos Pinto, "Spuren," 173–76.

12. Davis points out how the beginning of the organized women's movement is "often assumed to have its origins" in conversations that took place at the 1840 World Anti-Slavery Convention. Davis, *Women, Race & Class*, 40. In the 1980s, feminist organizations in Switzerland began fighting for migrants and thematizing racism because of the influence from US Black feminists especially, who critiqued the absence of discussing racism in women's movements. Similarly, Audre Lorde's visits to Germany spurred organizing among Black women during this time. Burke et al., *I Will Be Different Every Time*, 57.

13. Dos Santos Pinto, "Spuren," 173.

14. With the heading "What Voting Is For," I am referencing partly Malcolm X's 1964 speech "The Ballot or the Bullet" because of how he outlines in it what voting and political activism *mean* for Black lives: that voting must be turned into activism, a way to fight for Black lives. Malcolm X, "The Ballot or the Bullet."

15. Murakawa elaborates on the ways in which racism is hidden in plain sight by a placing of racism into and onto a certain set of overt expressions and acts of discriminations. Murakawa, *The First Civil Right*, 7–8.

16. Swiss Confederation, "Federal Constitution of the Swiss Confederation," art. 10a.
17. Dos Santos Pinto et al., *Un/Doing Race*, 9.
18. Eskandari and Banfi have addressed the interplay between the notion of Swiss customs, or "traditional values," and Islamophobic rhetoric during the campaigns for this law in detail. Eskandari and Banfi, "Institutionalising Islamophobia in Switzerland."
19. Dos Santos Pinto et al., *Un/Doing Race*, 9.
20. Dos Santos Pinto et al., *Un/Doing Race*, 10.
21. Chandler, *"Beyond This Narrow Now,"* 9.
22. Chandler, *"Beyond This Narrow Now,"* 9.
23. Chandler, *"Beyond This Narrow Now,"* 9.
24. Leaning on Gordon, "Theory in Black."
25. Leaning on Gordon, "Theory in Black," 199.

CHAPTER 6. MOTHERSHIP CONNECTIONS
1. Castile, "For Mother of Philando Castile."
2. Nash, *Birthing Black Mothers*, 185.
3. Oguntoye et al., *Showing Our Colors*, 48–49.
4. Han, "Slavery as Contract," 395.
5. Han, "Slavery as Contract," 396, 408.
6. Han, "Slavery as Contract," 401.
7. See Moten, *Stolen Life*, 255.
8. Han, "Slavery as Contract," 408–9.
9. See Wilopo and Häberlein, "Illegalisierung und Race."
10. See Spillers, "Mama's Baby, Papa's Maybe."
11. Moten, *Stolen Life*, 260–61.
12. Leaning here on Moten, *Stolen Life*, 260.
13. Moten, *Stolen Life*, 260–61.
14. See Honohan, *The Theory and Politics of Ius Soli*, 3–5.
15. See Sesay, "The Revolutionary Black Roots of Slavery's Abolition in Massachusetts."
16. Stovall, *White Freedom*, 73.
17. Stovall, *White Freedom*, 320–21.
18. Sesay, "The Revolutionary Black Roots of Slavery's Abolition in Massachusetts," 109–10.
19. Spillers, "Mama's Baby, Papa's Maybe," 80.
20. Spillers asks Black men to embrace the female within, which becomes possible exactly because of this excess of mothering that is announced in Blackness's alternate kinship practices. Spillers, "Mama's Baby, Papa's Maybe," 80.
21. Brackets in citation. Irigaray, *This Sex Which Is Not One*, 170.
22. Irigaray, *This Sex Which Is Not One*, 170–71.
23. Irigaray, *This Sex Which Is Not One*, 172.
24. Irigaray, *This Sex Which Is Not One*, 174.
25. Irigaray, *This Sex Which Is Not One*, 174–75.
26. Harris, "Whiteness as Property."

27. Cox, "Stories of the Mothership," 117.
28. See also Schuller, *The Biopolitics of Feeling*.
29. See, for example, Amin, "Trans* Plasticity and the Ontology of Race and Species."
30. Ndiaye, "Rewriting the *Grand Siècle*," 4–8.
31. See chapter 2 for an extended analysis of this good and bad blood question.
32. See Bridgewater, *Breeding a Nation*; Morgan, *Laboring Women*; Smithers, *Slave Breeding*; Stanley, "Slave Breeding and Free Love."
33. Day, "Afro-Feminism before Afropessimism," 69.
34. Day, "Afro-Feminism before Afropessimism," 79.
35. Diprose and Ziarek, *Arendt, Natality and Biopolitics*, 129. For an account of Agamben's articulation of biopolitics' relation to women's reproductive rights, see Cerwonka and Loutfi, "Biopolitics and the Female Reproductive Body as the New Subject of Law," 3.
36. Diprose and Ziarek, *Arendt, Natality and Biopolitics*, 187.
37. Irigaray, *This Sex Which Is Not One*, 206.
38. Diprose and Ziarek, *Arendt, Natality and Biopolitics*, 199–23.
39. Ndikung, *The Delusions of Care*.
40. Vergès, *The Wombs of Women*, 60.
41. Ndiaye, *Scripts of Blackness*, 98.
42. J. Brown, "Being Cellular," 337.
43. Brackets in citation. Miller, *The Limits of Bodily Integrity*, 149.
44. Waldby and Cooper, "The Biopolitics of Reproduction," 58.
45. Goodwin, "No, Justice Alito, Reproductive Justice Is in the Constitution."
46. Goodwin, *Policing the Womb*, 11.
47. Diprose and Ziarek, *Arendt, Natality and Biopolitics*, 217.
48. Diprose and Ziarek, *Arendt, Natality and Biopolitics*, 5–6.
49. Benhabib, "Arendt's Eichmann in Jerusalem," 81–82.
50. Moten, *The Universal Machine*, 100.
51. Leaning on Moten, *Stolen Life*, 266.
52. See Strathern, *Relations*.
53. Spillers, "Mama's Baby, Papa's Maybe," 80.
54. Gumbs, "We Can Learn to Mother Ourselves," 48.

CHAPTER 7. INTERSTITIAL LISTENINGS IV

1. For a rumination on breathing and Black lives, see Cox and François, "Aesthetics of (Black) Breathing."
2. Culver and Hauck, "8 Minutes, 46 Seconds and 'Inherently Dangerous.'"
3. Forliti, "Prosecutors: Officer Had Knee on Floyd for 7:46, not 8:46."
4. Willis et al., "New Footage Shows Delayed Medical Response to George Floyd."
5. Uzor, *8′46″ George Floyd in Memoriam*.
6. As Whitesell proposes in his analysis of the racial aspects in some of Cage's avant-garde peers: "In these quests for the irreducible background we can see the ideals of 'art without ties' converging under the sign of racial identity. That is, the blank page comes to serve as medium of white self-representation." Whitesell, "White Noise," 175.

7. Davies, "John Cage's 4′33″," 449.
8. Dohoney, "John Cage, Julius Eastman, and the Homosexual Ego," 49–50.
9. See also Lewis, "Improvised Music after 1950," 99.
10. Davies, "John Cage's 4′33″," 449.
11. Refering to Lewis, "Lifting the Cone of Silence from Black Composers."
12. *Merriam-Webster*, s.v. "white noise."
13. ATIS is overseen by the American National Standards Institute (ANSI), an organization that determines standards of various kinds, such as products, terms, processes, systems, and more. ATIS, "Search Keyword By."
14. Wilderson, *Red, White & Black*, 80.
15. See also Thompson, "Whiteness and the Ontological Turn in Sound Studies."
16. There is a certain protest of silence that is heard as a noisy black, as protesting Blacks. See Stoever, *The Sonic Color Line*, 3.
17. See, for example, Wellman, "Black Squares for Black Lives?"; Chang et al., "#JusticeforGeorgeFloyd."
18. See, for example, Shaw, "Kasimir Malevich's *Black Square*," for whom the *Black Square* represents the absence of representation.
19. Black, "Fractal Freedoms," 5–6.
20. Tarrow, "What Does Malevich's 'Black Square' Mean?"
21. Shatskikh, *Black Square*, x.
22. Reinhardt in Rose, *Art-as-Art*, 87; Moten, "The Case of Blackness," 190–93.
23. After Moten, "The Case of Blackness."
24. While I don't hear Shatskikh necessarily saying such, her analysis makes this possible reading of Malevich's work evident as well as an alternative one too. Shatskikh, *Black Square*, 261–62, 273–74.
25. Leaning on Cox, "Cecil Taylor's Posthumanistic Musical Score."
26. Moten, "The Case of Blackness," 194.
27. See Marriott, "Judging Fanon," 20–22.
28. Moten, "The Case of Blackness," 191.
29. Moten, "The Case of Blackness," 195.
30. See Marriott, *Lacan Noir*, 58.
31. Leaning on Marriott's analysis of Lacan. Marriott, "Ontology and *Lalangue*," 231.
32. After Marriott, "Ontology and Lalangue," 231.
33. Moten, "The Case of Blackness," 194.
34. Leaning on Marriott's analysis of Lacan: "He [Lacan] misinterprets how white self-certainty makes blackness impossible because it cannot know it; he misinterprets the logic of mastery as a desire for recognition rather than arising from the fearful uncertainty regarding others, from which it derives; he misinterprets decision and desire because he is content to say what is certain or decidable is equivalent to a desire to be white." Marriott, "Ontology and *Lalangue*," 231.
35. Adu-Sanyah, "Current Work."
36. Adu-Sanyah, "About."
37. Black, "Fractal Freedoms," 8.

CHAPTER 8. LISTENING WITH BLACK SWITZERLAND

1. I'm referencing here particularly a kind of Afrofuturist respelling of blackness. See, for example, Akomfrah, *The Last Angel of History*; Cokes, "Filmmaker's Journal."
2. Thomas, *Froudacity*, 240–41.
3. Brooks beautifully elaborates on fugitive listening in "Fugitive Listening."
4. Crawford and Joler, "Anatomy of an AI System."
5. Crawford and Joler, "Anatomy of an AI System."
6. Sovacool, "The Precarious Political Economy of Cobalt."
7. Crawford and Joler, "Anatomy of an AI System."
8. Derrida, "Archive Fever," 10.
9. It works through a repetition that adds nothing but promises to add something once it is added. Derrida, "Archive Fever," 14.
10. As Derrida points out in "Archive Fever," but see also Foucault, *The Archeology of Knowledge*, 145–46.
11. How Du Bois's "Sociology Hesitant" is a question of narrative refiguring, see Judy, "Introduction," 35.
12. Derrida, "Archive Fever," 12.
13. Derrida, "Archive Fever," 54.
14. I'm respelling here Derrida's observation that the archive is "spectral." Derrida, "Archive Fever," 54.
15. Hall, "Constituting an Archive," 91.
16. Parliament, "Mothership Connection (Star Child)."
17. Spillers, "Mama's Baby, Papa's Maybe," 72.
18. Spillers, "Mama's Baby, Papa's Maybe," 67.
19. Weheliye, *Habeas Viscus*, 39.
20. We might also say, as a question of what Fanon identifies as "epidermalization."
21. Spillers, "Mama's Baby, Papa's Maybe," 67.
22. Something along the lines of Derridean *grammatology*. Derrida, *Of Grammatology*. For an account of how language and meaning is encyclopedic, see Eco, "Dictionary vs. Encyclopedia."
23. For an account of how listening is entangled in racialization, see Stoever, *The Sonic Color Line*.
24. Stoever, *The Sonic Color Line*, 7, 8, 14–15.
25. Since listening is of the archive, it functions always through and with a deposition, a placing that is both material (geographical and biological) and symbolical. That is also to say that an account of the materiality of sound or listening does not necessarily disarticulate an antiblack gaze. Material as distinct domain, over which archival deposition takes authority, is authority's validation through ownership of fleshly parts, of making resources into parts. For example, Chung elaborates on how theories that attempt to claim an absence of the symbolical and to purely follow matter can lead to a flat ontology, which could be in such critique likened to the critiques of colorblindness and multiculturalism. Chung, "Vibration, Difference, and Solidarity in the Anthropocene."

26. For an elaboration of listening's structural enmeshedness, see Robinson, *Hungry Listening*.

27. Gordon makes a similar point regarding disciplines and methodology in "Theory in Black," 201–2.

28. Spivak, "Can the Subaltern Speak?," 70.

29. Spivak, "Can the Subaltern Speak?," 70.

30. Du Bois, *Black Reconstruction in America*, 41.

31. Morris, "Sociology of Race and W. E. B. DuBois," 504; McKee, *Sociology and the Race Problem*.

32. Judy, *(Dis)forming the American Canon*, 92; Wilderson, *Red, White & Black*, 41.

33. I, of course, also refer here to the notion of natalpolitics as elaborated in chapter 6. See also Goodwin, *Policing the Womb*, 41–43.

34. Chandler, "Originary Displacement," 165–66.

35. Hall, "Constituting an Archive," 92.

36. *Improvisation*, in my use of the term, is not simply a question of variation and reperformance but rather defined by this scene of fugitivity—it inaugurates something that's impossible in and outside of the world as the excess of repetition, or the lack of the possibility of such.

37. See Cox, "Stories of the Mothership," 117.

38. Glissant, *Introduction to a Poetics of Diversity*, 8; Cox and Yulsman, "Listening through Webs for/of Creole Improvisation," 3.

39. Moten, "Blackness and Nothingness," 762.

40. Derrida, *Sovereignties in Question*, 41.

41. Derrida bespeaks here the chain of signifiers as well as desire. Derrida, *Sovereignties in Question*, 41.

42. I might speculate fugitive librarian practices as such after Mattern, "Fugitive Libraries."

43. Cox, "Stories of the Mothership."

44. Derrida, "Force of Law," 971.

45. Leaning on Althusser, "Idéologie et appareils idéologiques d'État."

CHAPTER 9. INTERSTITIAL LISTENINGS V

1. See Sharpe, *In the Wake*.

2. For a detailed analysis of listening's investment in the reproduction of race, see Stoever, *The Sonic Color Line*.

3. Browne, *Dark Matters*, 113–14; Dyer, *White*, 103.

4. Lewis, "The Racial Bias Built into Photography."

5. Deleuze and Guattari, *A Thousand Plateaus*, 348, as quoted in Barrett, "Between Noise and Language," 159.

6. Deleuze and Guattari, *A Thousand Plateaus*, 348, as quoted in Barrett, "Between Noise and Language," 159.

7. Monk, "Criss Cross."

8. Adorno, "Music, Language, and Composition," 405.

9. Chandler, "Originary Displacement," 165–66.

10. I'm referring here to Gordon's elaboration of theory as illicit seeing, which is equatable with looking at blackness. Gordon, "Theory in Black," 196–97.

11. Chandler, "Originary Displacement," 165–66.

12. Sun Ra and His Arkestra, "The Alter Destiny."

13. Sharpe, *In the Wake*, 17.

14. Gordon, "Theory in Black."

15. Wilderson, *Red, White & Black*, 82.

16. Blow, "The Great Erasure."

CHAPTER 10. BLACK LIFE / SCHWARZ-SEIN

1. Coltrane, *Universal Consciousness*.

2. Derrida bespeaks this, albeit as a question of the sign, in Sigmund Freud in "Archive Fever," 61–62.

3. After Derrida, "Shibboleth."

4. After W. E. B. Du Bois.

5. Weheliye, "Black Life / Schwarz-Sein," 254.

6. Weheliye, "Black Life / Schwarz-Sein," 253–54.

7. Weheliye, "Black Life / Schwarz-Sein," 254.

8. Mackey, "Other."

9. Weheliye, "Black Life / Schwarz-Sein," 254.

10. Weheliye, "Black Life / Schwarz-Sein," 255.

11. Warren, *Ontological Terror*, 6. Strikethrough in the original.

12. Barad, "What Is the Measure of Nothingness?," 17.

13. Barad, "What Is the Measure of Nothingness?," 15.

14. Sun Ra, *Nothing Is. . . .*

15. Copeland highlights such in relation to Barad's work in "Tending-Toward-Blackness."

16. What I called "uncertainty" in Cox, "Stories of the Mothership."

17. Their name came from abolitionist "Harriet Tubman's 1853 raid on the Combahee River in South Carolina that freed 750 enslaved people." Taylor, *How We Get Free*, 4.

18. Combahee River Collective, "The Combahee River Collective Statement," 266–67.

19. Taylor, *How We Get Free*, 9.

20. Sexton, "Basic Black."

21. The philosophers are Martin Heidegger, Jean-Paul Sartre, and Friedrich Wilhelm Joseph Schelling. Leung, "Nothingness without Reserve."

22. Leung, "Nothingness without Reserve," 9.

23. Marriott, "Corpsing," 33–34.

24. Glissant and Delpech-Hellsten, "Rien n'est vrai, tout est vivant," 1–2.

25. And I am here also thinking of the kind of alter-destinies proposed by the Black radical tradition as elaborated by, for example, J. Brown, *Black Utopias*.

26. Campt, "Afro-German Cultural Identity and the Politics of Positionality," 120.

27. Dos Santos Pinto, "Tilo Frey und die nichtperformative Inklusion," 66–68.

28. Tages-Anzeiger, "Doppelte Emanzipation für die Schweiz," 5, as cited in dos Santos Pinto, "Tilo Frey und die nichtperformative Inklusion," 70.

29. Dos Santos Pinto, "Tilo Frey und die nichtperformative Inklusion," 70–71.

30. "Entrée Féminine Fleurie Sous La Coupole Fédérale," *Feuilles D'Avis De Neuchâtel*, 1.

31. Campt discusses this in relation to Black German Hans Hauck's acceptance by the Nazis. Campt, *Other Germans*, 111.

32. Nenno, "Reading the 'Schwarz' in the 'Schwarz-Rot-Gold,'" 3.

33. See Walcott, "Outside in Black Studies."

34. And here I mean this term *Black*, which bespeaks the very uprooting of antiblackness, the refiguring of lives, what life can be, what kinship can mean, performed as a kind of nonperformance, as an avant-garde kind of exceeding-the-stage performance that cannot be contained or delimited by any categories, names, or terms authorized by an antiblack world.

35. Wilderson, *Red, White & Black*, 38.

36. Slum Village, "Fall in Love."

37. Moten, *The Universal Machine*, 167.

38. Ramshaw elaborates on how improvisation is of the impossible and in her re-reading of Derrida's account of improvisation—who evokes it as always already failed—presents a *petition* for a radical togetherness and refiguring of justice. She re-presents the meeting between Ornette Coleman and Derrida, and in this shifts something of Derrida's resistance to falling into improvisation, where failure becomes something else—we might say, the impossible that is always here. Or, in other words, the impossible as failure of improvisation, as improvisations failing of being present, is revealed in Derrida, through a re-reading with Ramshaw, and possibly also this book, which is also to say, definitely, with Black life, as not the lack of presence but presentness's nonwholeness by itself. Improvisation as impossible work, or failure as impossible's encroachment on the world, is a radical justice transfiguration that comes forward in a care for each other. Thus, if Black life marks failure of the world, and such failure is the impossible improvisation out of this world, then the impossible is the improvisational reforming of the world through blackness as excess of the bounds of any world in and by itself. Ramshaw, "Deconstructin(g) Jazz Improvisation," 8–9.

39. After Jenkins, "Unspoken Grammar of Place."

40. Chandler, *"Beyond This Narrow Now,"* 185.

41. Chandler, *"Beyond This Narrow Now,"* 224.

42. Glissant, *Poetics of Relation*, 5–9.

43. Leaning on Moten, "Jurisgenerative Grammar (for Alto)."

44. Chandler, *"Beyond This Narrow Now,"* 225.

45. In this passage I am writing after the elaboration of the question of the case, Heidegger, Fanon, and black squares, among other things, in Moten, "The Case of Blackness."

46. Which is what Heidegger implies in "Das Ding," 176.

47. Or social death.

48. Griffin, *Read Until You Understand*.

49. Moten, "The Subprime and the Beautiful," 240.

50. Leaning on Ossie and the Mystic Revelation of Rastafari, *Grounation*.

CONCLUSION. ALONGSIDE A CHORUS OF VOICES

1. And a team behind the scenes.

2. The Lucerne Festival (of 2022) lineup had already been curated before the curation of Lucerne Festival Forward 2021 had been completed. The theme was already decided upon in 2019. Haefliger, "Festival-Thema 'Diversity.'" While Baureithel und Berg mention that the Forward Festival has given up more power without turning to a motto like "Diversity," which was seen as problematic by them, it in fact is not so simple. While a more "democratic" approach does have the potential to widen the pool of potential people to be curated, antiblackness can still persist and the structures of exclusion can be reproduced again, as we've uncovered in this book. Baureithel and Berg, "Wie weiss muss klassische Musik sein?"

3. Lewis, "A Small Act of Curation," 16.

4. Ensemble Modern, "Afromodernism in Contemporary Music."

5. Color-blindness as analyzed in chapter 2. At the Afromodernism conference, Björn Gottstein, one of the foremost curators, who was just, during that time, wrapping up his tenure as curator for Donaueschingen (one of the most important new music festivals), spoke on how paradoxical it was that a field that sees itself as forward thinking had never confronted this problematic within their own community.

6. Through devices such as slave collars with bells.

7. Lewis, *A Power Stronger Than Itself*.

8. Moten, *Stolen Life*, 241.

9. Referencing the Afrofuturists in Eshun, "Further Considerations of Afrofuturism," 287.

10. Halberstam, "The Wild Beyond," 8.

11. Spivak, "Critical Intimacy."

12. Fanon, *Black Skin, White Masks*, 17, as quoted in Harney and Moten, *The Undercommons*, 96.

13. Harney and Moten, *The Undercommons*, 96–97.

14. I hear this in the possibility of affirmation elaborated by Sexton, who hears it after Marriott and also in Moten's work. Sexton, "Ante-Anti-Blackness."

15. We might call this a form of disappearance, as Black Swiss denotes here the meeting of nothingness and existence. In this sense, it exceeds the very structure of recognition that is tied to an antiblack world. After Fanon, *Wretched of the Earth*, 246; Chari, "Exceeding Recognition," 119.

16. Moten, "Notes on Passage," 51.

17. Walia, *Undoing Border Imperialism*, 5–6.

18. See Walia, *Border and Rule*, 3.

19. See Barad, "Troubling Time/s and Ecologies of Nothingness."

POSTFACE. ENDLESS ENDLESSNESS

1. Frei, "Seine Musik macht das Sterben hörbar."
2. Frei, "Seine Musik macht das Sterben hörbar."
3. Frei, "Seine Musik macht das Sterben hörbar."
4. Martinot and Sexton, "The Avant-Garde of White Supremacy."
5. Martinot and Sexton, "The Avant-Garde of White Supremacy," 179.
6. Frei, "Seine Musik macht das Sterben hörbar."
7. Hartman, *Scenes of Subjection*, 4.
8. Hartman, *Scenes of Subjection*, 4; Moten, *Black and Blur*, viii.
9. As Sexton critiques: "That is to say, the reiteration and circulation of passages like Douglass's description of Aunt Hester's torture do not help to establish for readers 'the violent tenor of the slave's life,' as Mbembe would have it, but rather serve to obscure and to naturalize it." Sexton, "People-of-Color-Blindness," 34. But we must move beyond this critique, for it fails to account for the desire to mark this event as exception, which antiblackness uses, particularly in Europe, to erase the event of antiblackness through containing antiblack brutalities in an event that can be localized in the past or elsewhere, like the United States.
10. After Moten, *Black and Blur*, ix–x.
11. Leaning on Moten, *Stolen Life*, 17, 27, 33, 35.
12. SWI swissinfo.ch, "Court Acquits Swiss Police of Death of Nigerian Man."
13. Juillard, "Rapports accablants."
14. After Sun Ra, *The Immeasurable Equation*, 74.

Bibliography

Ablinger, Peter. "Quadraturen." Ablinger.mur.at (website). Last updated August 21, 2006. https://ablinger.mur.at/docu11.html.

Abraham, William. "The Life and Times of Anton Wilhelm Amo." *Transactions of the Historical Society of Ghana* 7 (1964): 60–81.

Adorno, Theodor W. "Music, Language, and Composition." Translated by Susan Gillespie. *Musical Quarterly* 77, no. 3 (1993): 401–14.

"Ad Reinhardt Abstract Painting 1960–1961." Museum of Modern Art. Accessed October 11, 2023. https://www.moma.org/collection/works/79982.

Adu-Sanyah, Akosua Viktoria. "About." Akosuaviktoria.com. Accessed March 23, 2023. https://akosuaviktoria.com/about.

Adu-Sanyah, Akosua Viktoria. "Current Work." Akosuaviktoria.com. Accessed March 23, 2023. https://akosuaviktoria.com/INHERITANCE.

AFP / The Local. "Protests Force Switzerland to Grapple with Its Own 'George Floyd' Case." *Local*, June 16, 2020. https://www.thelocal.ch/20200616/switzerland-grapples-with-its-own-george-floyd-case.

Agamben, Giorgio. *Homo Sacer: Sovereign Power and Bare Life*. Stanford, CA: Stanford University Press, 1998.

Agamben, Giorgio. *Remnants of Auschwitz: The Witness and the Archive*. New York: Zone, 1999.

Ahmed, Sara. "Declarations of Whiteness: The Non-Performativity of Anti-Racism." *Borderlands* 3, no. 2 (2004). https://webarchive.nla.gov.au/awa/20050616083826/http://www.borderlandsejournal.adelaide.edu.au/vol3no2_2004/ahmed_declarations.htm.

Ahmed, Sara. "The Nonperformativity of Antiracism." *Meridians* 7, no. 1 (2006): 104–26.

Akomfrah, John, dir. *The Last Angel of History*. London: Black Audio Film Collective, 1996.

Alliance against Racial Profiling. "The Alliance." Accessed August 20, 2022. https://www.stop-racial-profiling.ch/en/allianz-gegen-racial-profiling.

Althusser, Louis. "Idéologie et appareils idéologiques d'État." *La Pensée* 151 (1970): 3–38.

Amin, Kadji. "Trans* Plasticity and the Ontology of Race and Species." *Social Text* 38, no. 2 (2020): 49–71.

ATIS. "Search Keyword By: Black Noise." Alliance for Telecommunications Industry Solutions. Accessed March 24, 2023. https://glossary.atis.org/search-results/?search=black+noise.

Ayim, May. *Grenzenlos und unverschämt*. Berlin: Orlanda Frauenverlag, 1997.

Barad, Karen. *Meeting the Universe Halfway*. Durham, NC: Duke University Press, 2007.

Barad, Karen. "Troubling Time/s and Ecologies of Nothingness: Re-turning, Re-membering, and Facing the Incalculable." *New Formations* 92, no. 92 (2017): 56–86.

Barad, Karen. "What Is the Measure of Nothingness? Infinity, Virtuality, Justice." In *100 Notes, 100 Thoughts: Documenta Series 099*. Kassel: Hatje Cantz, 2012.

Barrett, G. Douglas. "Between Noise and Language: The Sound Installations and Music of Peter Ablinger." *Mosaic: An Interdisciplinary Critical Journal* (2009): 147–64.

Batumike, Chikuru. *Être Noir africain en Suisse*. Paris: L'Harmattan, 2006.

Batumike, Cikuru. *Noirs de Suisse*. Nice: Les Éditions Ovadia, 2014.

Batumike, Cikuru. *Présence africaine en Suisse*. Paris: La Pensée Universelle, 1993.

Baureithel, Elisabeth, and Jenny Berg. "Wie weiss muss klassische Musik sein?" SRF Swiss Radio and Television, August 10, 2022. https://www.srf.ch/kultur/musik/diversitaet-am-lucerne-festival-wie-weiss-muss-klassische-musik-sein.

Benhabib, Seyla. "Arendt's Eichmann in Jerusalem." In *The Cambridge Companion to Hannah Arendt*, edited by Dana Richard Villa, 65–85. Cambridge: Cambridge University Press, 2000.

Bernasconi, Robert. "Who Invented the Concept of Race?" In *Theories of Race and Racism: A Reader*, 2nd ed., edited by Les Back and John Solomos, 83–103. London: Routledge, 2009.

Beyoncé. "Me, Myself and I." On *Dangerously in Love*. Columbia COL 509395 2, 2003.

Black, Hannah. "Fractal Freedoms." *Afterall: A Journal of Art, Context and Enquiry* 41, no. 1 (2016): 4–9.

Black Central Europe. "Home." Accessed March 23, 2023. https://blackcentraleurope.com.

Blow, Charles M. "The Great Erasure." *New York Times*, May 20, 2022.

Bonaventure. "Supremacy." On *Free Lutangu*. Purple Tape Pedigree, 2003, 2017.

Bonilla-Silva, Eduardo. *Racism without Racists: Color-Blind Racism and the Persistence of Racial Inequality in the United States*. Oxford: Rowman and Littlefield, 2006.

Borio, Gianmario. "Work Structure and Musical Representation: Reflections on Adorno's Analyses for Interpretation." *Contemporary Music Review* 26, no. 1 (2007): 53–75.

Boulila, Stefanie Claudine. "Race and Racial Denial in Switzerland." *Ethnic and Racial Studies* 42, no. 9 (2019): 1401–18.

Bratt, Christopher. "Is It Racism? The Belief in Cultural Superiority across Europe." *European Societies* 24, no. 2 (2022): 207–28.

Bridgewater, Pamela D. *Breeding a Nation: Reproductive Slavery, the Thirteenth Amendment, and the Pursuit of Freedom*. Brooklyn: South End Press, 2014.

Broad, David B. "The Social Register: Directory of America's Upper Class." *Sociological Spectrum* 16, no. 2 (1996): 173–81.

Brooks, Andrew Navin. "Fugitive Listening: Sounds from the Undercommons." *Theory, Culture & Society* 37, no. 6 (2020): 25–45.

Brown, Jayna. "Being Cellular: Race, the Inhuman, and the Plasticity of Life." *GLQ: A Journal of Lesbian and Gay Studies* 21, no. 2–3 (2015): 321–41.

Brown, Jayna. *Black Utopias: Speculative Life and the Music of Other Worlds*. Durham, NC: Duke University Press, 2021.

Brown, Wesley. *Darktown Strutters: A Novel*. New York: Cane Hill, 1994.

Browne, Simone. *Dark Matters: On the Surveillance of Blackness*. Durham, NC: Duke University Press, 2015.

Bruggmann, Elvie, Marie Dixon Seidel, Carmel Fröhlicher-Stines, Verena Hillmann, Chantal-Nina Kouoh, Sabina Larcher, Rita Imelda Volkart, and Susi Wiederkehr, eds. "Draussen—drinnen—dazwischen: Women of Black Heritage." *Olympe: Feministische Arbeitshefte zur Politik* 18, no. 3 (2003).

Bundi, Sabrina. "Schwarze Schweizer über Rassismus: 'Mir hat mal einer von hinten einfach in die Haare gefasst.'" *Tages Anzeiger*, February 5, 2023. https://www.tagesanzeiger.ch/wie-schwarz-duerfen-schweizerinnen-und-schweizer-sein-278598905421.

Burke, Fork, Myriam Diarra, and Franziska Schutzbach. *I Will Be Different Every Time: Schwarze Frauen in Biel / Femmes Noires à Biennes / Black Women in Biel*. Biel/Bienne: Verlag die Brotsuppe, 2020.

Busey, Christopher L., and Tianna Dowie-Chin. "The Making of Global Black Anti-Citizen/Citizenship: Situating BlackCrit in Global Citizenship Research and Theory." *Theory & Research in Social Education* 49, no. 2 (2021): 153–75.

Campt, Tina M. "Afro-German Cultural Identity and the Politics of Positionality: Contests and Contexts in the Formation of a German Ethnic Identity." *New German Critique*, no. 58 (1993): 109–26.

Campt, Tina Marie. *Other Germans: Black Germans and the Politics of Race, Gender, and Memory in the Third Reich*. Ann Arbor: University of Michigan Press, 2009.

"Carmel Fröhlicher-Stines." Histnoire.ch (website). Accessed August 6, 2022. https://histnoire.ch/material/carmel-froehlicher-stines.

Castile, Valerie. "For Mother of Philando Castile." Interview by Reid Forgrave, *Star Tribune*, June 4, 2020. https://www.startribune.com/for-mother-of-philando-castile-george-floyd-s-death-a-nightmare-revisited/570997652.

Cerwonka, Allaine, and Anna Loutfi. "Biopolitics and the Female Reproductive Body as the New Subject of Law." *feminists@law* 1, no. 1 (2011).

Chandler, Nahum Dimitri. *"Beyond This Narrow Now": Or, Delimitations, of W. E. B. Du Bois*. Durham, NC: Duke University Press, 2021.

Chandler, Nahum Dimitri. "Originary Displacement: or, Passages of the Double and the Limit of World." In *X—The Problem of the Negro as a Problem for Thought*, 129–70. New York: Fordham University Press, 2013.

Chandler, Nahum Dimitri. *Toward an African Future—of the Limit of World*. Albany: State University of New York Press, 2021.

Chang Ho-Chun Herbert, Allissa Richardson, and Emilio Ferrara. "#Justicefor-GeorgeFloyd: How Instagram Facilitated the 2020 Black Lives Matter Protests." *PLOS ONE* 17, no. 12 (2022): e0277864. https://doi.org/10.1371/journal.pone.0277864.

Chari, Anita. "Exceeding Recognition." *Sartre Studies International* 10, no. 2 (2004): 110–22.

Chénière, Maïté. *Sonic S.cape*. Digital. Accessed August 13, 2022. http://www.11.digital/show/octavial-scape-maite-cheniere.

Chung, Andrew J. "Vibration, Difference, and Solidarity in the Anthropocene: Ethical Difficulties of New Materialist Sound Studies and Some Alternatives." *Resonance: The Journal of Sound and Culture* 2, no. 2 (2021): 218–41.

Cokes, Tony. "Filmmaker's Journal: *Resonanz. 01* (2008– 2013) Notes/Fragments on a Case of Sonic Hauntology." *Black Camera* 5, no. 1 (2013): 220–25.

Collectif Afro-Swiss. "Cas quoi . . . ?" Accessed March 30, 2023. https://collectifafroswiss.wordpress.com/6-2/.

Coltrane, Alice. *Universal Consciousness*. Impulse! Records, 1971.

Combahee River Collective. "The Combahee River Collective Statement." In *Home Girls: A Black Feminist Anthology*, edited by Barbara Smith, 264–74. New Brunswick, NJ, and London: Rutgers University Press, 2000.

Coney, John, dir. *Space Is the Place*. Los Angeles: Jim Newman, 1974.

Conseil Municipal. "Mémorial." *La Ville de Genève*. Accessed April 24, 2024. https://conseil-municipal.geneve.ch/conseil-municipal/memorial.

Cooperaxion. "Datenbank der im Sklavenhandel involvierten Schweizer." Accessed March 24, 2023. https://cooperaxion.org/sklavenhandel.

Copeland, Huey. "Tending-Toward-Blackness." *October* 156 (2016): 141–44.

Coulthard, Glen Sean. *Red Skin, White Masks: Rejecting the Colonial Politics of Recognition*. Minneapolis: University of Minnesota Press, 2014.

Cox, Jessie. "Cecil Taylor's Posthumanistic Musical Score." *American Music Review* 50, no. 2 (2021). https://www.brooklyn.cuny.edu/web/aca_centers_hitchcock/AMR_50-2_Cox.pdf.

Cox, Jessie. "Stories of the Mothership: Charles Uzor and *Mothertongue*." In *Composing While Black: Afrodiasporische Neue Musik Heute/Afrodiasporic New Music Today*, edited by Harald Kisiedu and George E. Lewis, 114–29. Hofheim: Wolke-Verlag, 2023.

Cox, Jessie, and Isaac Jean-François. "Aesthetics of (Black) Breathing." *liquid blackness* 6, no. 1 (2022): 98–117.

Cox, Jessie, and Sam Yulsman. "Listening through Webs for/of Creole Improvisation: Weaving Music II as a Case Study." *Critical Studies in Improvisation / Études critiques en improvisation* 14, no. 2–3 (2021): 1–13.

Crawford, Kate, and Vladan Joler. "Anatomy of an AI System: The Amazon Echo as an Anatomical Map of Human Labor, Data and Planetary Resources." *AI Now Institute and Share Lab* 7 (2018).

Cretton, Viviane. "Performing Whiteness: Racism, Skin Colour, and Identity in Western Switzerland." *Ethnic and Racial Studies* 41, no. 5 (2017): 842–59.

Culver, Jordan, and Grace Hauck. "8 Minutes, 46 Seconds and 'Inherently Dangerous': What's in the Criminal Complaint in the George Floyd Case?" *USA Today*, May 29, 2020.

Dancy, Christopher L., and P. Khalil Saucier. "AI and Blackness: Toward Moving beyond Bias and Representation." *IEEE Transactions on Technology and Society* 3, no. 1 (2021): 31–40.

Davies, Stephen. "John Cage's 4'33": Is it Music?" *Australasian Journal of Philosophy* 75, no. 4 (1997): 448–62.

Davis, Angela Y. *Women, Race & Class*. London: Penguin Classics, 2019. First published by Random House, 1981.

Davis, Miles. *In a Silent Way*. Columbia CS 9875, 1969.

Day, Iyko. "Afro-Feminism before Afropessimism: Meditations on Gender and Ontology." In *Antiblackness*, edited by Moon-Kie Jungand and João H. Costa Vargas, 60–81. Durham, NC: Duke University Press, 2021.

Degonda, Silvana. "Rapper Nativ macht sich gegen Rassismus stark: 'Ich werde immer wieder von der Polizei kontrolliert.'" *Schweizer Illustrierte*, July 6, 2020. https://www.schweizer-illustrierte.ch/people/swiss-stars/rapper-nativ-ich-werde-immer-wieder-von-der-polizei-kontrolliert-304031.

Deleuze, Gilles, and Félix Guattari. *A Thousand Plateaus: Capitalism and Schizophrenia*. Translated by Brian Massumi. Minneapolis: University of Minnesota Press, 1987.

Derrida, Jacques. "Archive Fever: A Freudian Impression." Translated by Eric Prenowitz. *Diactrics* 25, no. 2 (1995): 9–63.

Derrida, Jacques. "Force of Law: The 'Mystical Foundation of Authority.'" Translated by Mary Quaintance. *Cardozo Law Review* 11, no. 5–6 (1990): 920–1046.

Derrida, Jacques. *Of Grammatology*. Translated by Gayatri Chakravorty Spivak. Baltimore, MD: Johns Hopkins University Press, 1976.

Derrida, Jacques. "Shibboleth: For Paul Celan." In Derrida, *Sovereignties in Question*, 1–64.

Derrida, Jacques. *Sovereignties in Question: The Poetics of Paul Celan*. Edited by Thomas Dutoit and Outi Pasanen. New York: Fordham University Press, 2005.

"Des Moutons pour sensibiliser aux valeurs du 'Vivre Ensemble.'" *Vivre à Genève*, no. 24, December (2007): 19.

Diprose, Rosalyn, and Ewa Plonowska Ziarek. *Arendt, Natality and Biopolitics: Toward Democratic Plurality and Reproductive Justice*. Edinburgh: Edinburgh University Press, 2018.

Dohoney, Ryan. "John Cage, Julius Eastman, and the Homosexual Ego." *Tomorrow Is the Question: New Directions in Experimental Music Studies* (2014): 39–62.

dos Santos Pinto, Jovita. "Spuren. Eine Geschichte Schwarzer Frauen in der Schweiz." In *Terra Incognita? Der Treffpunkt Schwarzer Frauen in Zürich*, edited by Shelley Berlowitz, Elisabeth Joris, and Zeedah Meierhofer-Mangeli, 143–85. Zurich: Limmat Verlag, 2013.

dos Santos Pinto, Jovita. "Tilo Frey und die nichtperformative Inklusion." In *Un/Doing Race: Rassifizierung in der Schweiz*, edited by Jovita dos Santos Pinto, Pamela Ohene-Nyako, Mélanie-Evely Pétrémont, Anne Lavanchy, Barbara Lüthi, Patricia Purtschert, and Damir Skenderovic, 55–75. Zurich: Seismo Verlag, 2022.

dos Santos Pinto, Jovita, Pamela Ohene-Nyako, Mélanie-Evely Pétrémont, Anne Lavanchy, Barbara Lüthi, Patricia Purtschert, and Damir Skenderovic, eds. *Un/Doing Race: Rassifizierung in der Schweiz*. Zurich: Seismo Verlag, 2022.

"Dr. Mutombo Kanyana." Schauspielhaus Zürich. Accessed March 30, 2023. https://www.schauspielhaus.ch/de/personen/23033/dr-mutombo-kanyana.

Du Bois, W. E. B. *Black Reconstruction in America*. New York: Harcourt, Brace, 1935.

Du Bois, W. E. B. "My Evolving Program for Negro Freedom." *Clinical Sociology Review* 8, no. 1 (1990): 27–57.

Du Bois, W. E. B. *The Philadelphia Negro*. New York: Cosimo, 2007. First published 1899.

Du Bois, W. E. B. "Sociology Hesitant." *boundary 2* 27, no. 3 (2000): 37–44.

Du Bois, W. E. B. *Writings: The Suppression of the Slave-Trade / The Souls of Black Folk / Dusk of Dawn / Essays and Articles*. Edited by Nathan Huggins. New York: Library of America, 1986.

Dyer, Richard. *White: Essays on Race and Culture*. New York: Routledge, 1997.

Eco, Umberto. "Dictionary vs. Encyclopedia." In *Semiotics and the Philosophy of Language*. Bloomington: Indiana University Press, 1986.

El-Tayeb, Fatima. *European Others: Queering Ethnicity in Postnational Europe*. Minneapolis: University of Minnesota Press, 2011.

El-Tayeb, Fatima. *Schwarze Deutsche: Der Diskurs um "Rasse" und Nationale Identität 1890–1933*. Frankfurt: Campus Verlag, 2001.

Ensemble Modern. "Afromodernism in Contemporary Music." May 11, 2022. https://www.ensemble-modern.com/en/news/2020-11-05/afro-modernism-in-contemporary-music.

"Ensemble of the Lucerne Festival Contemporary Orchestra (LFCO)." Lucerne Festival. Accessed October 11, 2023. https://www.lucernefestival.ch/en/program/ensemble-of-the-lucerne-festival-contemporary-orchestra-lfco-mariano-chiacchiarini-sofia-jernberg-winnie-huang/2007.

"Entrée féminine fleurie sous la Coupole Fédérale." *Feuilles d'Avis de Neuchâtel*, November 30, 1971. https://www.e-newspaperarchives.ch/?a=d&d=EXR19711130-01&e=fr-20-1-img-txIN-0.

Eshun, Kodwo. "Further Considerations of Afrofuturism." *CR: The New Centennial Review* 3, no. 2 (2003): 287–302.

Eskandari, Vista, and Elisa Banfi. "Institutionalising Islamophobia in Switzerland: The Burqa and Minaret Bans." *Islamophobia Studies Journal* 20, no. 10 (2017): 53–71.

"Expertengruppe zu Rassismus: Schweiz weist Rassismus-Vorwürfe der UNO zurück." SRF Schweizer Radio und Fernsehen, March 10, 2022. https://www.srf.ch/news/schweiz/expertengruppe-zu-rassismus-schweiz-weist-rassismus-vorwuerfe-der-uno-zurueck.

Fanon, Frantz. *Black Skin, White Masks*. Translated by Charles Lam Markmann. New York: Grove, 2008.

Fanon, Frantz. *The Wretched of the Earth*. Translated by Constance Farrington. New York: Grove, 1963.

Farred, Grant. "Daseinstufe: Liquidity as a (Distinct) Stage of Being." *liquid blackness* 5, no. 1 (2021): 39–61.

Fernandez, Xavier. "Affaire Mike Ben Peter: 'Je n'ai jamais été confronté à d'autres cas où la personne arrêtait de respirer.'" *Le Matin*, June 12, 2023. https://www.lematin.ch/story/je-n-ai-jamais-ete-confronte-a-d-autres-cas-ou-la-personne-arretait-de-respirer-609867516175.

Forliti, Amy. "Prosecutors: Officer Had Knee on Floyd for 7:46, not 8:46." AP News, June 17, 2020. https://apnews.com/article/0b4714f6a42b362b0e2c0cd701c6392b.

Forschungskollektiv Rassismus vor Gericht. "Racial Profiling vor Gericht—Der Fall Mohamed Wa Baile: Setzen Sie sich weiterhin gegen Diskriminierung ein, aber folgen Sie den Anweisungen der Polizei." Alliance against Racial Profiling, February 2017. https://www.stop-racial-profiling.ch/en/court_cases/mo-wa-baile.

Foucault, Michel. *The Archeology of Knowledge*. Translated by A. M. Sheridan Smith. New York: Routledge, 2002. First published by Editions Gallimard, 1969.

Foucault, Michel. "What Is an Author?" In *Aesthetics: A Reader in Philosophy of the Arts*, edited by David Goldblatt, Lee B. Brown, and Stephanie Patridge, 284–88. New York: Routledge, 2017.

Frei, Marco. "Seine Musik macht das Sterben hörbar." *Neue Zürcher Zeitung*, November 15, 2023. https://www.nzz.ch/feuilleton/lucerne-festival-forward-darf-kunst-das-sterben-eines-menschen-instrumentalisieren-ld.1765529.

Fröhlicher-Stines, Carmel. "The Effects of Racism on Group and Individual Identity." *Schatten der Vergangenheit und die Last der Bilder. Rassismus gegen Schwarze in der Schweiz*, Nationale Tagung der EKR, March 20, 2002. https://www.ekr.admin.ch/pdf/020320_tagung_carmel_froehlicher_stines_en11a3.pdf.

Fröhlicher-Stines, Carmel, and Kelechi Monika Mennel. *Schwarze Menschen in der Schweiz: Ein Leben zwischen Integration und Diskriminierung*. Bern: EKR/CFR, 2004.

Garcia, Megan. "Racist in the Machine." *World Policy Journal* 33, no. 4 (2016): 111–17.

Gilroy, Paul. *The Black Atlantic: Modernity and Double Consciousness*. Cambridge, MA: Harvard University Press, 1993.

Glissant, Édouard. "From the Whole-World Treatise." Translated by A. James Arnold. *Review: Literature and Arts of the Americas* 32, no. 58 (1999): 31–34.

Glissant, Édouard. *Introduction to a Poetics of Diversity*. Vol. 1. Translated by Celia Britton. Liverpool: Liverpool University Press, 2020.

Glissant, Édouard. *Poetics of Relation*. Translated by Betsy Wing. Ann Arbor: University of Michigan Press, 1997.

Glissant, Édouard, and Catherine Delpech-Hellsten. "Rien n'est vrai, tout est vivant." Translated by Alexandre Leupin. *Glissant Translation Project* (blog), October 20, 2017. https://sites01.lsu.edu/wp/theglissanttranslationproject/2017/10/20/nothing-is-true-everything-is-living.

Goldstein, Sidney. "City Directories as Sources of Migration Data." *American Journal of Sociology* 60, no. 2 (1954): 169–76.

Gondola, Didier. "'But I Ain't African, I'm American!' Black American Exiles and the Construction of Racial Identities in Twentieth-Century France." In *Blackening Europe*, edited by Paul Gilroy and Heike Raphael-Hernandez, 201–15. New York: Routledge, 2012.

Goodwin, Michele. "No, Justice Alito, Reproductive Justice Is in the Constitution." *New York Times*, June 26, 2022.

Goodwin, Michele. *Policing the Womb: Invisible Women and the Criminalization of Motherhood*. Cambridge: Cambridge University Press, 2020.

Gordon, Lewis R. *Bad Faith and Antiblack Racism: A Study in the Philosophy of Jean-Paul Sartre*. New Haven, CT: Yale University Press, 1993.

Gordon, Lewis R. "Theory in Black: Teleological Suspensions in Philosophy of Culture." *Qui Parle: Critical Humanities and Social Sciences* 18, no. 2 (2010): 193–214.

Griffin, Farah Jasmine. *Read Until You Understand: The Profound Wisdom of Black Life and Literature*. New York: W. W. Norton, 2021.

Gumbs, Alexis Pauline. *Undrowned: Black Feminist Lessons from Marine Mammals*. San Francisco: AK, 2020.

Gumbs, Alexis Pauline. "We Can Learn to Mother Ourselves: The Queer Survival of Black Feminism 1968–1996." PhD diss., Duke University, 2010.

Haefliger, Michael. "Festival-Thema 'Diversity.'" Interview by Urs Mattenberger, *Piu: Das Magazin zum Lucerne Festival* (Summer 2022): 6–7.

Halberstam, Jack. "The Wild Beyond." In *The Undercommons: Fugitive Planning and Black Study*, by Stefano Harney and Fred Moten, 2–13. Brooklyn: Automedia / Minor Compositions, 2013.

Hall, Stuart. "Constituting an Archive." *Third Text* 15, no. 54 (2001): 89–92.

Han, Sora. "Slavery as Contract: Betty's Case and the Question of Freedom." *Law & Literature* 27, no. 3 (2015): 395–416.

Hannaford, Marc Edward. "One Line, Many Views: Perspectives on Music Theory, Composition, and Improvisation through the Work of Muhal Richard Abrams." PhD diss., Columbia University, 2019.

Harney, Stefano, and Fred Moten. *The Undercommons: Fugitive Planning and Black Study*. Brooklyn: Automedia / Minor Compositions, 2013.

Harris, Cheryl I. "Whiteness as Property." *Harvard Law Review* (1993): 1707–91.

Hartman, Saidiya. *Lose Your Mother: A Journey along the Atlantic Slave Route*. New York: Farrar, Straus and Giroux, 2007.

Hartman, Saidiya V. *Scenes of Subjection: Terror, Slavery, and Self-making in Nineteenth-Century America*. Oxford: Oxford University Press, 1997.

Hartman, Saidiya. "Venus in Two Acts." *Small Axe* 12, no. 2 (2008): 1–14.

Heidegger, Martin. "Das Ding (1950)." *Gesamtausgabe I. Abteilung: Veröffentliche Schriften 1910–1976, Band 7, Vorträge und Aufsätze*. Frankfurt: Vittorio Klostermann, 2000.

Hérnandez, Javier C. "A European Music Festival's Push for Diversity Stirs Debate." *New York Times*, August 12, 2022.

Hill, Lawrence. *The Book of Negroes: A Novel*. New York: W. W. Norton, 2009.

Honohan, Iseult. *The Theory and Politics of Ius Soli*. Robert Schuman Centre for Advanced Studies, in Collaboration with Edinburgh University Law School Comparative Report, 2010.

hooks, bell. "Eating the Other: Desire and Resistance." In *Black Looks: Race and Representation*, 21–39. Boston: South End, 1992.

hooks, bell. *Talking Back: Thinking Feminist, Thinking Black*. Boston: South End, 1989.
Hull, David L. "Are Species Really Individuals?" *Systematic Zoology* 25, no. 2 (1976): 174–91.
Irigaray, Luce. *This Sex Which Is Not One*. Ithaca, NY: Cornell University Press, 1985.
Irigaray, Luce. "Women on the Market." In *This Sex Which Is Not One*. Ithaca, NY: Cornell University Press, 1985.
Jenkins, DeMarcus A. "Unspoken Grammar of Place: Anti-Blackness as a Spatial Imaginary in Education." *Journal of School Leadership* 31, no. 1–2 (2021): 107–26.
Jolo, Jérémie. "Als Kind wünschte ich mir eine Zeit lang, weiss zu werden." Interview by Beat Kuhn, *Bieler Tagblatt*, January 10, 2022.
Jolo, Jérémie. "Perspectives and Blurred Colors—Piece for Clarinet, Voice and Loopstation." YouTube, March 23, 2022. https://www.youtube.com/watch?v=6iTnoGopYzc.
Jolo, Jérémie. "RED, 2018." YouTube, January 27, 2021. https://www.youtube.com/watch?v=FU45svXiiO8.
"Jovita dos Santos Pinto über Alltagsrassismus." SRF Schweizer Radio und Fernsehen, May 3, 2019. https://www.srf.ch/kultur/gesellschaft-religion/black-lives-matter-hoert-uns-zu-informiert-euch-und-unterstuetzt-uns.
Judy, Ronald A. *(Dis)forming the American Canon: African-Arabic Slave Narratives and the Vernacular*. Minneapolis: University of Minnesota Press, 1993.
Judy, Ronald A. "Introduction: On WEB Du Bois and Hyperbolic Thinking." *boundary 2* 27, no. 3 (2000): 1–35.
Judy, Ronald A. "The Unfungible Flow of Liquid Blackness." *liquid blackness* 5, no. 1 (2021): 27–36.
Juillard, Amit. "Rapports accablants: Selon des experts Américains, Mike Ben Peter est mort à cause de la police." *Blick*, November 26, 2023. https://www.blick.ch/fr/news/suisse/rapports-accablants-selon-des-experts-americains-mike-ben-peter-est-mort-a-cause-de-la-police-id19166412.html.
Jung, Moon-Kie, and João H. Costa Vargas, eds. *Antiblackness*. Durham, NC: Duke University Press, 2021.
Kay, Jean. "Berlin-Based Producer Bonaventure on Trauma, White Supremacy + Music as an Act of War." AQNB, February 23, 2017. https://www.aqnb.com/2017/02/23/free-lutangu-berlin-producer-bonaventure-on-trauma-white-supremacy-the-act-of-war.
Khan, Stephen. "Dehumanisation, Animalisation: Inside the Terrible World of Swiss Human Zoos." *Conversation*, June 22, 2023. https://theconversation.com/dehumanisation-animalisation-inside-the-terrible-world-of-swiss-human-zoos-208133#.
Krampe, Christian J. *The Past Is Present: The African-Canadian Experience in Lawrence Hill's Fiction*. New York: Peter Lang, 2012.
Kreis, Georg. "Bundesgericht entlastet SVP-Politiker: Nur ein bisschen rassistisch." *WOZ: Die Wochenzeitung*, September 27, 2012. https://www.woz.ch/1239/bundesgericht-entlastet-svp-politiker/nur-ein-bisschen-rassistisch.
Kukuruz Quartet and Julius Eastman. *Piano Interpretations*. Zurich: Intakt Records, 2018.

Kusmierz, Zoë Antonia. "Areas of Uncertainty: Observations on the German Reception of Spike Lee and Hip-Hop Culture." In *Amerikanische Populärkultur in Deutschland: Case Studies in Cultural Transfer Past and Present*, edited by Heike Paul and Katja Kanzler, 167–80. Leipzig: Leipziger Universitätsverlag, 2002.

L'Abri. "Maïté Chénière." Accessed August 13, 2022. https://labrigeneve.ch/en/maite-cheniere.

Lavanchy, Anne. *How Does "Race" Matter in Switzerland?* Neuchâtel: Université de Neuchâtel, 2014.

Layne, Priscilla Dionne. *White Rebels in Black: German Appropriation of Black Popular Culture*. Ann Arbor: University of Michigan Press, 2018.

"Le Parti Radical répond aux affiches polémiques de l'UDC montrant un mouton noir 2007." RTS Radio Télévision Suisse. Accessed March 24, 2023. http://www.rts.ch/video/info/journal-19h30/1490083-le-parti-radical-repond-aux-affiches-polemiques-de-ludc-montrant-un-mouton-noir.html.

Leung, King-Ho. "Nothingness without Reserve: Fred Moten contra Heidegger, Sartre, and Schelling." *Comparative and Continental Philosophy* (2022): 1–13.

Lewis, George E. "Improvised Music after 1950: Afrological and Eurological Perspectives." *Black Music Research Journal* (1996): 91–122.

Lewis, George E. "Lifting the Cone of Silence from Black Composers." *New York Times*, July 3, 2020.

Lewis, George E. *A Power Stronger Than Itself: The AACM and American Experimental Music*. Chicago: University of Chicago Press, 2008.

Lewis, George E. "A Small Act of Curation." *OnCuring* 44 (2019). https://on-curating.org/issue-44-reader/a-small-act-of-curation.html#.Zi1YbnbMI7c.

Lewis, Sarah. "The Racial Bias Built into Photography." *New York Times*, April 25, 2019.

L'Harmattan. "Cikuru Batumike." Editions L'Harmattan (website). Accessed October 11, 2023. https://www.editions-harmattan.fr/index.asp?navig=auteurs&obj=artiste&no=12363.

Logan, Olivia. "The American Poet Who Started the Afro-German Movement." *Exberliner*, November 24, 2020. https://www.exberliner.com/politics/lorde-afro-german.

"Logothetis 100.3." Wien Modern. Accessed July 30, 2022. https://www.wienmodern.at/2021-logothetis-1003-20211104-0800-reaktor-en.

Lusane, Clarence. *Hitler's Black Victims: The Historical Experiences of Afro-Germans, European Blacks, Africans, and African Americans in the Nazi Era*. New York: Routledge, 2003.

Lüthi, Barbara, Francesca Falk, and Patricia Purtschert. "Colonialism without Colonies: Examining Blank Spaces in Colonial Studies." *National Identities* 18, no. 1 (2015): 1–9. https://doi.org/10.1080/14608944.2016.1107178.

Mackey, Nathaniel. "Other: From Noun to Verb." In *The Improvisation Studies Reader: Spontaneous Acts*, edited by Ajay Heble and Rebecca Caines, 261–78. New York: Routledge, 2014.

Maddocks, Fiona. "Lucerne Festival 2022 Review." *Guardian*, September 10, 2022.

Malcolm X. "The Ballot or the Bullet." In *Malcolm X Speaks: Selected Speeches and Statements*, edited by George Breitman, 23–44. London: Secker and Warburg, 1966.

Marriott, David. "Corpsing; or, The Matter of Black Life." *Cultural Critique* 94 (2016): 32–64.

Marriott, David. "Judging Fanon." *Rhizomes: Cultural Studies in Emerging Knowledge* 29, no. 1 (2016). https://doi.org/10.20415/rhiz/029.e03.

Marriott, David. *Lacan Noir: Lacan and Afropessimism*. Cham: Palgrave Macmillan, 2021.

Marriott, David S. "Ontology and *Lalangue* (or, Blackness and Language)." *Critical Philosophy of Race* 10, no. 2 (2022): 220–47.

Martinot, Steve, and Jared Sexton. "The Avant-Garde of White Supremacy." *Social Identities* 9, no. 2 (2003): 169–81.

Mattern, Shannon. "Fugitive Libraries." *Places Journal* (October 2019).

Mbembé, Achille. *Critique of Black Reason*. Durham, NC: Duke University Press, 2017.

Mbembé, Achille. "Necropolitics." Translated by Libby Meintjes. *Public Culture* 15, no. 1 (2003): 11–40.

Mbembé, Achille. *Necropolitics*. Translated by Steven Corcoran. Durham, NC: Duke University Press, 2019.

McKee, James B. *Sociology and the Race Problem: The Failure of a Perspective*. Urbana: University of Illinois Press, 1993.

Merriam-Webster. S.v. "white noise. (*n*.)." Accessed March 24, 2023. https://www.merriam-webster.com/dictionary/white%20noise.

Meyer, Mattea. "Prävention gegen die Diskriminierung bei Personenkontrollen durch das Grenzwachkorps." Das Schweizer Parlament. Accessed October 11, 2023. https://www.parlament.ch/de/ratsbetrieb/suche-curia-vista/geschaeft?AffairId=20183353.

Michel, Noémi. "Racial Profiling und die Tabuisierung von 'Rasse.'" In *Un/Doing Race: Rassifizierung in der Schweiz*, edited by Jovita dos Santos Pinto, Pamela Ohene-Nyako, Mélanie-Evely Pétrémont, Anne Lavanchy, Barbara Lüthi, Patricia Purtschert, and Damir Skenderovic, 101–19. Zurich: Seismo Verlag, 2022.

Michel, Noémi. "Sheepology: The Postcolonial Politics of Raceless Racism in Switzerland." *Postcolonial Studies* 18, no. 4 (2015): 410–26.

Miller, Ruth A. *The Limits of Bodily Integrity: Abortion, Adultery, and Rape Legislation in Comparative Perspective*. London: Routledge, 2016.

Monk, Thelonious. "Criss Cross." On *Criss Cross*, Columbia CS 8838, 1963.

Morgan, Jennifer. *Laboring Women: Reproduction and Gender in New World Slavery*. Philadelphia: University of Pennsylvania Press, 2004.

Morris, Aldon D. "Sociology of Race and W. E. B. DuBois: The Path Not Taken." In *Sociology in America*, edited by Craig Calhoun, 503–34. Chicago: University of Chicago Press, 2008.

Morrison, Matthew D. "Race, Blacksound, and the (Re)Making of Musicological Discourse." *Journal of the American Musicological Society* 72, no. 3 (2019): 781–823.

Morrison, Matthew D. "Sound in the Construction of Race: From Blackface to Blacksound in Nineteenth-Century America." PhD diss., Columbia University, 2014.

Moten, Fred. *Black and Blur*. Vol. 1. Durham, NC: Duke University Press, 2017.

Moten, Fred. "Blackness and Nothingness (Mysticism in the Flesh)." *South Atlantic Quarterly* 112, no. 4 (2013): 737–80.

Moten, Fred. "The Case of Blackness." *Criticism* 50, no. 2 (2008): 177–218.

Moten, Fred. *In the Break: The Aesthetics of the Black Radical Tradition*. Minneapolis: University of Minnesota Press, 2003.

Moten, Fred. "Jurisgenerative Grammar (for Alto)." In *The Oxford Handbook of Critical Improvisation Studies*, vol. 1, edited by George E. Lewis and Benjamin Piekut, 128–42. New York: Oxford University Press, 2016.

Moten, Fred. "Notes on Passage (the New International of Sovereign Feelings)." *Palimpsest* 3, no. 1 (2014): 51–74.

Moten, Fred. "Not in Between." In *Black and Blur*, vol. 1, 1–27. Durham, NC: Duke University Press, 2017.

Moten, Fred. *Stolen Life*. Vol. 2. Durham, NC: Duke University Press, 2018.

Moten, Fred. "The Subprime and the Beautiful." *African Identities* 11, no. 2 (2013): 237–45.

Moten, Fred. *The Universal Machine*. Vol. 3. Durham, NC: Duke University Press, 2018.

Muñoz, José Esteban. *Disidentifications: Queers of Color and the Performance of Politics*. Minneapolis: University of Minnesota Press, 1999.

Murakawa, Naomi. *The First Civil Right: How Liberals Built Prison America*. New York: Oxford University Press, 2014.

Muvumbi, Abril K. "Black Lives Matter in Italy." *European Journal of Women's Studies* 30, no. 1 (2023): 31S–33S.

Nadel, Alan. *Invisible Criticism: Ralph Ellison and the American Canon*. Iowa City: University of Iowa Press, 1988.

Naguib, Tarek. "Mit Recht gegen Rassismus im Recht: Rechtsverfahren als Mittel des Widerstands." In *Racial Profiling: Struktureller Rassismus and antirassistischer Wiederstand*, edited by Mohamed Wa Baile, Serena O. Dankwa, Tarek Naguib, Patricia Purtschert, and Sarah Schilliger, 257–73. Bielefeld: Transcript Verlag, 2019.

Nash, Jennifer C. *Birthing Black Mothers*. Durham, NC: Duke University Press, 2021.

National Archives. "Book of Negroes." Sir Guy Carleton Papers, no. 10427, Nova Scotia Archives. Accessed March 24, 2023. https://archives.novascotia.ca/africanns/archives/?ID=26.

Nativ. "Diese Revolution beginnt in unseren Köpfen: Ein Gespräch with Rapper Nativ." Interview by Ugur Gültekin. *WOZ: Die Wochenzeitung*, June 18, 2020.

Nativ. "Sira." On *Baobab*. HRD REC, 2018.

Ndiaye, Noémie. "Rewriting the Grand Siècle: Blackface in Early Modern France and the Historiography of Race." *Literature Compass* 18, no. 10 (2021): e12603. https://doi.org/10.1111/lic3.12603.

Ndiaye, Noémie. *Scripts of Blackness: Early Modern Performance Culture and the Making of Race*. Philadelphia: University of Pennsylvania Press, 2022.

Ndikung, Bonaventure Soh Bejeng. *The Delusions of Care*. Berlin: Archive Books, 2021.

Nenno, Nancy P. "Reading the 'Schwarz' in the 'Schwarz-Rot-Gold': Black German Studies in the 21st Century." *Transit* 10, no. 2 (2016). https://doi.org/10.5070/T7102031168.

Nowatzki, Robert. "'Blackin' Up Is Us Doin' White Folks Doin' Us': Blackface Minstrelsy and Racial Performance in Contemporary American Fiction and Film." *Literature Interpretation Theory* 18, no. 2 (2007): 115–36.

Oguntoye, Katharina, May Ayim, and Dagmar Schultz, eds. *Showing Our Colors: Afro-German Women Speak Out*. Translated by Anne V. Adams, with Tina Campt, May Opitz, and Dagmar Schultz. Amherst: University of Massachusetts Press, 1992. Originally published as *Farbe bekennen: Afro-deutsche Frauen auf den Spuren ihrer Geschichte*. Berlin: Orlanda Frauenverlag, 1986.

Olaudah, Equiano. *The Interesting Narrative of the Life of Olaudah Equiano*. Edited by Angelo Costanzo. Ontario, Canada: Broadview Press, 2001.

Ossie, Count, and the Mystic Revelation of Rastafari. *Grounation*. New Dimension ND 0001–0006, 1973.

Parliament. "Mothership Connection (Star Child)." On *Mothership Connection*. Cassablanca NBLP 7022, 1975.

Patterson, Orlando. *Slavery and Social Death: A Comparative Study, with a New Preface*. Cambridge, MA: Harvard University Press, 2018.

Purtschert, Patricia. "The Return of the Native: Racialised Space, Colonial Debris and the Human Zoo." *Identities* 22, no. 4 (2015): 508–23.

Purtschert, Patricia, Francesca Falk, and Barbara Lüthi. "Switzerland and 'Colonialism without Colonies': Reflections on the Status of Colonial Outsiders." *Interventions* 18, no. 2 (2016): 286–302.

Purtschert, Patricia, and Harald Fischer-Tiné. *Colonial Switzerland: Rethinking Colonialism from the Margins*. London: Palgrave Macmillan, 2015.

Ramshaw, Sara. "Deconstructin(g) Jazz Improvisation: Derrida and the Law of the Singular Event." *Critical Studies in Improvisation/Études critiques en improvisation* 2, no. 1 (2006). https://doi.org/10.21083/csieci.v2i1.81.

Robinson, Dylan. *Hungry Listening: Resonant Theory for Indigenous Sound Studies*. Minneapolis: University of Minnesota Press, 2020.

Rose, Barbara, ed. *Art-as-Art: The Selected Writings of Ad Reinhardt*. New York: Viking, 1975.

Schinkel, Willem. "From Zoēpolitics to Biopolitics: Citizenship and the Construction of 'Society.'" *European Journal of Social Theory* 13, no. 2 (2010): 155–72.

Schmidt, Hannah. "Konkrete Probleme." VAN *Magazine*, August 24, 2022. https://van-magazin.de/mag/diversity-lucerne-festival.

Schmitt, Carl. *The Concept of the Political: Expanded Edition*. Translated by George Schwab. Chicago: University of Chicago Press, 2007. First published 1996.

Schmitt, Carl. *Political Theology: Four Chapters on the Concept of Sovereignty*. Translated by George Schwab. Chicago: University of Chicago Press, 2005.

Schuller, Kyla. *The Biopolitics of Feeling: Race, Sex, and Science in the Nineteenth Century*. Durham, NC: Duke University Press, 2018.

Schwarze Schweiz. "Archive." Accessed August 24, 2022. https://en.schwarzeschweiz.com/archive.

Sesay, Chernoh M., Jr. "The Revolutionary Black Roots of Slavery's Abolition in Massachusetts." *New England Quarterly* 87, no. 1 (2014): 99–131.

Sexton, Jared. *Amalgamation Schemes: Antiblackness and the Critique of Multiracialism*. Minneapolis: University of Minnesota Press, 2008.

Sexton, Jared. "Ante-Anti-Blackness: Afterthoughts." *Lateral* 1 (2012). https://csalateral.org/issue/1/ante-anti-blackness-afterthoughts-sexton.

Sexton, Jared. "Basic Black." *liquid blackness* 5, no. 2 (2021): 75–83.

Sexton, Jared. "People-of-Color-Blindness: Notes on the Afterlife of Slavery." *Social Text* 28, no. 2 (2010): 31–56.

Sharpe, Christina. *In the Wake: On Blackness and Being*. Durham, NC: Duke University Press, 2016.

Shatskikh, Aleksandra. *Black Square: Malevich and the Origin of Suprematism*. New Haven, CT: Yale University Press, 2012.

Shaw, Philip. "Kasimir Malevich's *Black Square*." In *The Art of the Sublime*, edited by Nigel Llewellyn and Christine Riding. London: Tate Research Publication, 2013. https://www.tate.org.uk/art/research-publications/the-sublime/philip-shaw-kasimir-malevichs-black-square-r1141459.

Simon, Patrick. "The Choice of Ignorance: The Debate on Ethnic and Racial Statistics in France." In *Social Statistics and Ethnic Diversity: Cross-National Perspectives in Classifications and Identity Politics*, edited by Patrick Simon, Victor Piché, and Amélie A. Gagnon, 65–87. London: Springer Open, 2015.

Singletary, Kimberly Alecia. "Everyday Matters: Haunting and the Black Diasporic Experience." In *Rethinking Black German Studies: Approaches, Interventions and Histories*, edited by Tiffany N. Florvil and Vanessa D. Plumly, 137–67. Oxford: Peter Lang, 2018.

Slum Village. "Fall in Love." On *Fantastic, Vol. 2*, Good Vibe Recordings GVR 2025-2, 2000.

Smithers, Gregory D. *Slave Breeding: Sex, Violence, and Memory in African American History*. Gainesville: University Press of Florida, 2012.

Sovacool, Benjamin K. "The Precarious Political Economy of Cobalt: Balancing Prosperity, Poverty, and Brutality in Artisanal and Industrial Mining in the Democratic Republic of the Congo." *Extractive Industries and Society* 6, no. 3 (July 2019): 915–39.

Spillers, Hortense J. "Mama's Baby, Papa's Maybe: An American Grammar Book." *Diacritics* 17, no. 2 (1987): 65–81.

Spivak, Gayatri C. "Can the Subaltern Speak?" In *Colonial Discourse and Post-colonial Theory*, edited by Patrick Williams and Laura Chrisman, 66–111. London: Routledge, 2015.

Spivak, Gayatri C. "Critical Intimacy: An Interview with Gayatri Chakravorty Spivak." Interview by Steve Paulson. *Los Angeles Review of Books*, July 26, 2019.

Stadler, Gustavus. "Never Heard Such a Thing: Lynching and Phonographic Modernity." *Social Text* 28, no. 1 (2010): 87–105.

Stadt Zürich Präsidialdepartement. "Stadttaler an Carmel Fröhlicher-Stines und an Zeedah Meierhofer-Mangeli." Stadt Zürich, March 21, 2023. https://www.stadt-zuerich.ch/prd/de/index/ueber_das_departement/medien/medienmitteilungen/2023/maerz/230321a.html.

Stanley, Amy Dru. "Slave Breeding and Free Love: An Antebellum Argument over Slavery, Capitalism, and Personhood." In *Capitalism Takes Command: The Social Transformation of Nineteenth-Century America*, edited by Michael Zakim and Gary J. Kornblith, 119–44. Chicago: University of Chicago Press, 2011.

Stevens, A. Rebecca. "Many Black People Get Killed by Police in Quiet Switzerland Too." Medium, April 1, 2021. https://medium.com/illumination-curated/many-black-people-get-killed-by-police-in-switzerland-too-99fa49bea1c3.

Stoever, Jennifer Lynn. *The Sonic Color Line: Race and the Cultural Politics of Listening.* Vol. 17. New York: New York University Press, 2016.

Stovall, Tyler. "The Color Line behind the Lines: Racial Violence in France during the Great War." *American Historical Review* 103, no. 3 (1998): 737–69.

Stovall, Tyler. *White Freedom: The Racial History of an Idea.* Princeton, NJ: Princeton University Press, 2022.

Strathern, Marilyn. *Relations: An Anthropological Account.* Durham, NC: Duke University Press, 2020.

Sun Ra. *The Immeasurable Equation: The Collected Poetry and Prose.* Edited by James L. Wolf and Harttnut Geerken. Wartaweil: WAITAWHILE, 2005.

Sun Ra. *Nothing Is. . . .* ESP-DISK 1045, 1966.

Sun Ra and His Arkestra. "The Alter Destiny." On *Greatest Hits: Easy Listening for Intergalactic Travel.* Evidence ECD 22219-2, 2000.

Sun Ra and His Arkestra. "Death Speaks to the Negro." On *Strange Worlds in My Mind.* Norton Records ED-365, 2010.

Sun Ra and the Year 2000 Myth Science Arkestra. "Face the Music." On *Live at the Hackney Empire.* Leo Records CD LR 214/215, 1994.

Sutherland, Tonia. "Making a Killing: On Race, Ritual, and (Re)Membering in Digital Culture." *Preservation, Digital Technology & Culture* 46, no. 1 (2017): 32–40.

Sutton, Philip. "Direct Me NYC 1786: A History of City Directories in the United States and New York City." New York Public Library, June 8, 2012. https://www.nypl.org/blog/2012/06/08/direct-me-1786-history-city-directories-US-NYC.

Swiss Confederation. "Federal Act on Foreign Nationals and Integration." Die Publikationsplattform des Bundesrechts, December 16, 2005. https://www.fedlex.admin.ch/eli/cc/2007/758/de#art_89.

Swiss Confederation. "Federal Act on Swiss Citizenship." Die Publikationsplattform des Bundesrechts, June 20, 2014. https://www.fedlex.admin.ch/eli/cc/2016/404/de.

Swiss Confederation. "Federal Constitution of the Swiss Confederation." Die Publikationsplattform des Bundesrechts, April 18, 1999. https://www.fedlex.admin.ch/eli/cc/1999/404/en.

Swiss Confederation. "Swiss Civil Code." Die Publikationsplattform des Bundesrechts, December 10, 1907. https://www.fedlex.admin.ch/eli/cc/24/233_245_233/en#book_1/tit_1/chap_1/lvl_A/lvl_I_V/lvl_1.

SWI swissinfo.ch. "Court Acquits Swiss Police of Death of Nigerian Man." Swiss Broadcasting Corporation, July 23, 2023. https://www.swissinfo.ch/eng/business/court-acquits-swiss-police-of-death-of-nigerian-man/48612552.

"Switzerland Must Urgently Confront Anti-Black Racism." United Nations, January 26, 2022. https://www.ohchr.org/en/press-releases/2022/01/switzerland-must-urgently-confront-anti-black-racism-un-experts.

Szwed, John. *Space Is the Place: The Lives and Times of Sun Ra.* Durham, NC: Duke University Press, 2020.

Tarrow, Samantha. "What Does Malevich's 'Black Square' Mean?" Artifex. Accessed March 23, 2023. https://artifex.ru/живопись/казимир-малевич/.

Taylor, Keeanga-Yamahtta, ed. *How We Get Free: Black Feminism and the Combahee River Collective.* Chicago: Haymarket, 2017.

Thomas, John Jacob. *Froudacity: West Indian Fables Explained.* Philadelphia: Gebbie, 1890.

Thompson, Marie. "Whiteness and the Ontological Turn in Sound Studies." *Parallax* 23, no. 3 (2017): 266–82.

Thurman, Kira. "Performing Lieder, Hearing Race: Debating Blackness, Whiteness, and German Identity in Interwar Central Europe." *Journal of the American Musicological Society* 72, no. 3 (2019): 825–65.

Thurman, Kira. *Singing like Germans: Black Musicians in the Land of Bach, Beethoven, and Brahms.* Ithaca, NY: Cornell University Press, 2021.

"To All Art Spaces in Switzerland / An die Kunsthäuser in der Schweiz / À tous les espaces d'art de Suisse / Cari spazi d'arte in Svizzera." Black Artists and Cultural Workers in Switzerland, June 9, 2021. https://blackartistsinswitzerland.noblogs.org.

Tunstall, Dwayne A. *Doing Philosophy Personally: Thinking about Metaphysics, Theism, and Antiblack Racism.* New York: Fordham University Press, 2013.

Université Populaire Africaine en Suisse. "Home." Accessed January 19, 2023. https://www.upaf.ch.

Uzor, Charles. *Bodycam Exhibit 3: George Floyd in Memoriam.* 2021.

Uzor, Charles. *8′46″ George Floyd in Memoriam.* June 2020.

Uzor, Charles. "Melody and the Phenomenology of Internal Time-Awareness." PhD diss., University of London, 2005.

Vergès, Françoise. *The Wombs of Women: Race, Capital, Feminism.* Durham, NC: Duke University Press, 2020.

Wa Baile, Mohamed, Serena O. Dankwa, Tarek Naguib, Patricia Purtschert, and Sarah Schilliger. *Racial Profiling: Struktureller Rassismus and antirassistischer Wiederstand.* Bielefeld: Transcript Verlag, 2019.

Walcott, Rinaldo. *The Long Emancipation: Moving toward Black Freedom.* Durham, NC: Duke University Press, 2021.

Walcott, Rinaldo. "Outside in Black Studies: Reading from a Queer Place in the Diaspora." In *Black Queer Studies: A Critical Anthology,* edited by E. Patrick Johnson and Mae G. Henderson, 90–105. Durham, NC: Duke University Press, 2005.

Waldby, Catherine, and Melinda Cooper. "The Biopolitics of Reproduction: Post-Fordist Biotechnology and Women's Clinical Labour." *Australian Feminist Studies* 23, no. 55 (2008): 57–73.

Walia, Harsha. *Border and Rule: Global Migration, Capitalism, and the Rise of Racist Nationalism.* Chicago: Haymarket, 2021.

Walia, Harsha. *Undoing Border Imperialism.* Chico, CA: AK, 2013.

Warren, Calvin L. *Ontological Terror: Blackness, Nihilism, and Emancipation.* Durham, NC: Duke University Press, 2018.

Weber, Samuel. "Bare Life and Life in General." *Grey Room* 46 (2012): 7–24.

Weheliye, Alexander G. "Black Life / Schwarz-Sein: Inhabitations of the Flesh." In *Beyond the Doctrine of Man,* edited by Joseph Drexler-Dreis and Kristien Justaert, 237–62. New York: Fordham University Press, 2019.

Weheliye, Alexander G. *Habeas Viscus: Racializing Assemblages, Biopolitics, and Black Feminist Theories of the Human*. Durham, NC: Duke University Press, 2014.

Wellman, Mariah L. "Black Squares for Black Lives? Performative Allyship as Credibility Maintenance for Social Media Influencers on Instagram." *Social Media + Society* 8, no. 1 (2022). https://doi.org/10.1177/20563051221080473.

Whitesell, Lloyd. "White Noise: Race and Erasure in the Cultural Avant-Garde." *American Music* 19, no. 2 (2001): 168–89.

Wilderson, Frank B., III. *Afropessimism*. New York: Liveright, 2020.

Wilderson, Frank B., III. *Incognegro: A Memoir of Exile and Apartheid*. Durham, NC: Duke University Press, 2015.

Wilderson, Frank B., III. *Red, White & Black: Cinema and the Structure of US Antagonisms*. Durham, NC: Duke University Press, 2010.

Willis, Haley, Evan Hill, Robin Stein, Christiaan Triebert, Ben Laffin, and Drew Jordan. "New Footage Shows Delayed Medical Response to George Floyd." *New York Times*, August 11, 2020. https://www.nytimes.com/2020/08/11/us/george-floyd-body-cam-full-video.html.

Wilopo, Claudia, and Jana Häberlein. "Illegalisierung und Race. Konturen einer rassismuskritischen Analyse der Situation von abgewiesenen Asylsuchenden in der Schweiz." In *Un/Doing Race: Rassifizierung in der Schweiz*, edited by Jovita dos Santos Pinto, Pamela Ohene-Nyako, Mélanie-Evely Pétrémont, Anne Lavanchy, Barbara Lüthi, Patricia Purtschert, and Damir Skenderovic, 77–100. Zurich: Seismo Verlag, 2022.

Wright, Michelle M. *Becoming Black: Creating Identity in the African Diaspora*. Durham, NC: Duke University Press, 2004.

Wright, Michelle M. "Others-from-Within from Without: Afro-German Subject Formation and the Challenge of a Counter-Discourse." *Callaloo* 26, no. 2 (2003): 296–305.

Wynter, Sylvia. "Towards the Sociogenic Principle: Fanon, Identity, the Puzzle of Conscious Experience, and What It Is Like to Be 'Black.'" In *National Identities and Sociopolitical Changes in Latin America*, edited by Mercedes Duran-Cogan and Antonio Gomez-Moriana, 30–66. London and New York: Routledge, 2001.

Wynter, Sylvia. "'We Know Where We Are From': The Politics of Black Culture from Myal to Marley." Paper presented at the Joint Meeting of the African Studies and the Latin American Studies Association, Houston, November 1977. Accessed January 17, 2023. https://monoskop.org/images/4/4a/Wynter_Sylvia_1977_We_Know_Where_We_Are_From_The_Politics_of_Black_Culture_from_Myal_to_Marley.pdf.

Yancy, George. *Look, a White! Philosophical Essays on Whiteness*. Philadelphia: Temple University Press, 2012.

Young, Christopher. "Rassismus for Gericht." *Jusletter*, September 18, 2017. https://jusletter.weblaw.ch/juslissues/2017/906/rassismus-vor-gerich_bf056abb17.html.

Zanni, Bettina. "Die Frage 'Woher kommst du?' ist rassistisch." 20 Minuten, June 19, 2020. https://www.20min.ch/story/die-frage-woher-kommst-du-ist-rassistisch-633722993313.

Index

Page numbers followed by *f* refer to figures.

Ablinger, Peter, 28, 162
abstraction, 28, 46, 119–20, 138, 162–64, 176; Blackness as, 112, 133–34, 160; grammar and, 157; identity and, 175; sound as, 162; woman's body as, 111
Adu-Sanyah, Akosua Viktoria, 11, 127–28, 131, 134–35, 203n2; *White Gaze II Black Square*, 11–12, 127, 128*f*, 131–32, 134–35
Agamben, Giorgio, 77, 79–82, 88, 113, 212n29, 213n35, 216n35. *See also* bare life; biopolitics; homo sacer; *Muselmann*; sovereignty; state of exception
Alongside a Chorus of Voices (Cox), 187–89, 191
antiblackness, 3–7, 9–13, 15, 21, 24, 26, 29, 35, 38, 43, 54–58, 66, 69, 71, 75, 98, 126, 158, 166, 182–83, 188, 190, 197–201, 204n17, 221n34, 222n2; artwork/artworld and, 134; black skin and, 40; border control and, 97; border policing and, 41, 144; in Canada, 67; citizenship and, 83–84, 111; erasure of, 18, 37, 180–81, 223n9; identification and, 143, 156; listening and, 157, 160, 163; mixedness and, 49; motherness and, 102; natality and, 117–18; new music and, 189; nothingness and, 174, 177; policing and, 90; politics and, 96; race and, 205n34; refusal of, 191; resistance to, 210n5; as right to ownership, 172; silence and, 153; in Switzer-

land, 33–34, 39, 42, 44–45, 51, 59–61, 65, 97, 193, 199–200; technologies and, 27–28; white gaze and, 76, 125, 179, 185; women's reproduction and, 112–14, 116
anti-citizenship, 83–84
archive, the, 19, 21–23, 28, 94–96, 140–51, 154–56, 162–63, 166, 168, 218n14; Blackness and, 21, 94, 152; listening and, 218n25
Arendt, Hannah, 84, 113, 116–18
art, 6, 12, 129–31, 133–35, 161–62, 199, 216n6; black, 152; institutions, 3; of masks, 146; musical, 28; spaces, 4; visual, 127–28
authenticity, 69, 86, 145
authority, 15, 23–24, 54–55, 74, 98, 109, 131, 177, 191; of antiblackness, 71, 86, 95, 144, 152, 157–58, 183, 185, 197–98, 200; archive and, 141–43, 145, 149–51, 154, 162–63, 166, 218n25; borders and, 72, 89, 119–20, 154, 157; citizenship and, 84, 107, 168; erasure and, 24; exceptionality and, 88; homo sacer and, 78–79, 106; identification and, 145; listening and, 140–41, 150, 155, 190; politics and, 85; proper life and, 87; slavery and, 9; sovereignty and, 77–79, 106, 166, 212n29; of the state, 103, 143, 193; white gaze and, 178–79, 183
Ayim, May, 56, 95

Barad, Karen, 61–62, 172, 177, 220n15
bare life, 79–80, 84, 87–88, 90, 113, 213n39

Barrett, Douglas G., 28, 162
Batumike, Cikuru, 6, 35, 55–56, 69, 204n17, 204n19, 209n83
Being, 170, 176; Black Being, 169–70; of Schwarz Sein, 171
belonging, 14, 36, 40, 46, 50–51, 56, 102–4, 106–7, 109, 135, 138, 146, 167, 171, 190, 203n2; biological, 106; citizenship and, 74–75, 83, 113, 194; identification and, 18, 61; uncertainty of, 31, 72; wealth and, 66. *See also* national belonging; proper belonging
Betty's Case, 103–7, 152, 182
biopolitics, 80–82, 84–86, 113–14, 213n39, 216n35. *See also* necropolitics
biopower, 85, 213n39
Black, Hannah, 128, 130–31, 135
Black Europeans, 2, 204n17
Black life, 7, 24, 91–92, 169–71, 185–86, 194, 210n14, 221n38; citizenship and, 57, 76; in Germany, 56; improvisations of, 184; silencing of, 126; wake work and, 166; white gaze and, 74. *See also* Schwarz-Sein
Black lives, 9–13, 16–17, 21–24, 26–31, 34–36, 41, 66, 69, 78, 94–96, 98, 130, 137, 140, 145, 160, 169, 181–82, 188–89, 194, 197, 205n32; alternative forms of kinship and, 105; antiblackness and, 4–5, 9, 14, 26, 39, 54, 58, 175, 183, 191, 200–201; archive and, 19; Atlantic Ocean and, 204n23; biases against, 36, 139; Black women and, 112, 120; border and, 68; breathing and, 121–22, 216n1; in Canada, 67; citizenship and, 84, 109, 168; control of, 30, 40, 144, 148, 159; death of, 21, 23; *8'46" George Floyd in Memoriam* (Uzor) and, 123, 126–27; erasure of, 5, 17–18, 24, 26, 29, 36, 46, 72, 81, 125, 149; Europe and, 56; freedom and, 107–8; in Germany, 5; identification and, 18, 30–31, 57, 71, 138, 143, 152–53, 158; killing of, 37, 199–200; musical pitches and, 166; necropoliticality and, 178; owning of, 111, 146; racism and, 180, 204n19; sound of, 2, 8, 38; stories of, 3, 5, 144, 167; study of, 5–7, 35, 40, 64–65, 143, 210n14; unidentifiableness of, 57–58; in the United States, 64, 68; voting and, 214n14; white gaze and, 74

Black lives in Switzerland, 2–3, 7–9, 17, 33, 40, 55, 65, 67, 98, 145, 203n5, 204n19, 206n2; identification and, 59; study of, 35, 39; politics and, 95; thematization of, 6
Black Lives Matter protests, 18, 72, 119, 166, 188–89; global, 2, 10–11, 17, 127; in Switzerland, 68–69
black noise, 123–27, 135, 152, 157–58
Black people, 18, 24, 54–56, 69, 107, 151, 205n36; Black life and, 169; Blackness and, 13, 194; ownership over, 71; police violence against, 199; registering, 39
black skin, 39–41, 148–49
Black studies, 5, 35, 68, 81, 181, 198, 204n21
black study, 3, 5, 35, 65, 138, 185–86; projects, 98, 181, 190–91, 194; in Switzerland, 6
Black Swiss, 2–3, 6–10, 16, 18, 24, 33–35, 55–56, 68, 137, 185, 197–98; belonging and, 190; citizenship and, 78; expelled from Switzerland, 104; experience, 69; identification and, 145, 167; as meeting of nothingness and existence, 222n15; racism and, 39, 180
Black Swiss lives, 2, 7, 9, 33–34, 138, 197; erasure of, 69; music of, 11; stories of, 24
Black women, 94–95, 112–15, 120, 174–75, 214n12
blood, 57, 85, 89, 141, 216n31; border policing and, 101; citizenship and, 47, 54 (*see also jus sanguinis*); improper, 138; impure, 112; kinship, 83; mixing, 49 (*see also* miscegenation); national, 116; relations, 48
Bonaventure, 4–5
border control, 29–30, 64, 71, 79, 86, 97, 109, 148–50, 154, 163; antiblack methods of, 57; blackness and, 45, 213n53; kin and, 47, 49; modern, 161; technology and, 41; violence of, 213n53
border patrols, 42, 53, 72, 78
border policing, 40–42, 101, 120, 125, 144, 146, 162; blackness and, 6, 72; citizenship and, 90; identification and, 36, 143; mixedness and, 49; technologies of, 139

borders, 6, 15, 18, 53, 56, 61, 67, 74, 94, 109, 118–20, 148–49, 153–54, 172, 177; antiblackness and, 200; of the archive, 143, 145, 163; Black lives and, 68; blackness and, 24, 41, 78, 98, 114, 120, 126, 159, 181, 183; citizenship and, 54, 89, 107, 119, 194; culture and, 97; erasure of, 84, 118, 157; identification and, 186; interstitial listening and, 28, 162–63; of life, 79, 114; listening and, 98, 138–39, 150, 163; management of, 141; music and, 165; national, 50, 108, 119, 155; national belonging and, 44; race and, 46, 49; the racial and, 42; of Swissness, 57; Switzerland's, 36, 45, 68; underground of, 72, 90, 120, 143, 174

border zones, 21, 28–30, 46, 48–49, 54, 68, 75, 98, 101–3, 109, 143–45, 154, 156–57, 162; bare life and, 80; Black lives and, 57; blackness and, 84, 89–90; citizenship and, 80, 83, 86, 89–90, 108; cultural practices as, 97; erasure of, 174; homo sacer and, 79; listening and, 150; music as, 164; natalpolitics and, 115; politics and, 85

Browne, Simone, 160–61

Busey, Christopher, 83–84

Cage, John, 122–23, 126, 130, 216n6; *4'33,"* 122–24, 126, 173–74

capital, 9–10, 75, 140

Chandler, Nahum Dimitri, 97, 153–54, 165, 184, 486, 210n15, 214n10

Chauvin, Derek, 19, 22, 122. *See also* Floyd, George; Uzor, Charles: *8'46" George Floyd in Memoriam*

Chénière, Maïté (Mighty), 91–92, 99; *Sonic S.cape*, 91–94

citizenship, 2, 10, 15, 24, 42, 46, 73, 75, 79–80, 83–87, 89–90, 96, 102–13, 117–19, 151, 166–68, 213n35; antiblackness and, 83–84; Black life and, 57, 76; blackness and, 41, 76, 86, 190; borders and, 97; gender and, 181; homo sacer and, 77–79; kinship and, 109–10; modern, 71, 89; proper, 89; sovereignty and, 143, 145, 166; Swiss, 36, 47–48, 50, 54–55, 59, 78, 104–6, 192–94; white gaze and, 74. *See also* anti-citizenship; noncitizenship

civil society, 29, 38, 79, 86–87, 112; antiblackness and, 57; assumed goodness of, 115; proper life and, 79; state of exception and, 81; white gaze and, 73–74; women's wombs and, 146

classical music, 10–11; whiteness and, 205n29

color-blindness, 6, 38–39, 42–43, 45–46, 49–51, 53, 174, 199; antiblack, 79; four frames of, 207n22; white gaze and, 74

color line, 51–52, 54, 83

Combahee River Collective, 174–75, 220n17

commodity, 156; black as, 75; flesh as, 147–48; value of, 14, 111; women as, 111, 193

community, 45–47, 175, 205n28, 222n5; American women's, 109; Black Swiss, 9; building, 120; ethical, 117; multicultural, 49; Swiss, 43, 45

corpsing, 177–79, 183

Crawford, Kate, 139–40

creolization, 15, 155

Cretton, Vivian, 50–53

critical fabulation, 21–22, 94

death, 8–9, 12, 14–15, 24, 82, 84–89, 97–98, 115, 118, 121, 142, 170–71, 178, 190–91; Black, 23–24, 165; Black life and, 171; of Black lives, 21, 23; blackness and, 185; borders and, 72; canon, 10; of death, 177; death-worlds, 90, 179; of the (name of the) father, 141–42; self and, 135; social, 14, 85, 171, 222n47

democracy, 65, 97–98; direct, 96

Democratic Republic of the Congo, 6, 140

Derrida, Jacques, 140–41, 156–58, 173, 218n10, 218n14, 219n41, 220n2, 221n38. *See also* archive, the

Diprose, Rosalyn, 113–14, 116

discrimination, 36, 42, 54, 181, 214n15

disidentification, 61, 64, 90, 212n26

diversity, 45, 117, 164, 188; of origins, 43, 49

dos Santos Pinto, Jovita, 94–95, 180–81

double consciousness, 135, 153, 169, 177, 192, 211n11

Dowie-Chin, Tianna, 83–84

Du Bois, W. E. B., 51, 64–68, 95–98, 144, 151, 157, 184, 214n10; city directory metaphor, 95, 155, 186; Negro and, 204n18; "Sociology Hesitant," 142–43, 218n11. *See also* double consciousness

INDEX ■ 245

Eastman, Julius, 62, 123
Echo (Amazon), 141–44
emancipation, 29, 147–48, 151–52, 158, 189–90; double, 181; of mankind, 65; narrative of, 165
Ensemble Modern, 10, 189
enslavement, 103, 110, 147. *See also* slavery
epidermalization, 211n6, 218n20
equality, 38, 53, 111, 125; sovereign, 127
Equiano, Olaudah, 153, 165–66
erasure, 23, 27, 69, 114; antiblackness and, 13, 18, 37, 180, 199; of the archive, 145; bare life and, 79–80, 88, 113; of biological kinship, 106–7; of Black, 201; of Black lives, 5, 18, 24, 36, 125; of blackness, 6, 24, 34, 56, 78–79, 89, 180–81; of Black skin, 40–41; of Black Swiss, 7; of borders, 118, 157; of border zone, 174; citizenship and, 84, 90, 109; of doubling, 163; of flesh, 148; narrative and, 30; performance and, 178–79; of race, 37, 45–46, 49–50, 57, 89; of racialization, 36, 148; of racism, 37, 45, 50, 207n25; of silence, 153, 173; of the unknown, 144; of women, 181
ethnicity, 41–42, 48, 66, 81
Europeanness, 10, 82
exchange, 14, 75, 104, 163, 171; of commodities, 112; value, 162, 176, 183; women and, 110–11, 115
exploitation, 10, 83, 194

failure, 177–79, 183, 186, 206n4, 221n38; citizenship and, 107; death as, 24; opacity and, 60; of performance, 182–83; of sovereignty, 105
Fanon, Frantz, 11, 72–76, 78, 131, 171, 192, 212n22, 221n45; *Black Skin, White Masks*, 72, 204n24, 205n30, 210n86, 211n11, 212nn19–20. *See also* epidermalization
fidelity, 26–28, 160, 162. *See also* realness
flesh, 94, 145–50, 153–55; hieroglyphics of the, 82, 147–50, 156
Floyd, George, 11, 123, 125, 152, 166, 198; murder of, 17–19, 23, 25, 29, 101, 122, 166, 199. *See also* Uzor, Charles: *8'46" George Floyd in Memoriam*; Uzor, Charles: *Bodycam Exhibit 3: George Floyd in Memoriam*

foreigners, 18, 36, 44–46, 48, 51–52, 80, 98, 168; blackness and, 54; Black Swiss as, 2; displacements of, 193; identification and, 138
foreignness, 18, 36
Foucault, Michel, 80–82, 84, 113, 144, 218n10. *See also* biopolitics
Fourteenth Amendment, 65, 107–8, 116
freedom, 47, 87, 97–98, 107–8, 130, 149, 184, 190; displacement and, 193; Equiano and, 153, 165–66; legal, 105–6; modernity and, 189; slavery and, 29, 103–4, 189
Frey, Tilo, 94, 180–81, 182*f*
Fröhlicher-Stines, Carmel, 39–40

gender, 112, 180–81, 204n16
gendering, 146, 181
Glissant, Édouard, 60, 63, 155, 178–79, 184, 214n5; *tout-monde*, 214n4
Germany, 52, 56, 102, 180, 203n3; Afro-Germany, 204n21; Black lives in, 5, 205n28, 206n2; Black studies in, 181; Black women in, 214n12; Frankfurt, 10

Hall, Stuart, 144, 154
Harney, Stefano, 191–92
Hartman, Saidiya, 21, 57, 171, 199–200. *See also* critical fabulation
Heidegger, Martin, 170, 220n21, 221nn45–46. *See also* Being
hereness, 56, 172, 174, 193
homo sacer, 77–80, 104, 106

identification, 18, 20, 30–31, 36–37, 53, 57, 59, 61, 65–66, 71–72, 75, 78, 109, 138, 141, 145, 148–49, 152–53, 158–60, 167, 176, 182, 185–86; antiblack methods of, 16; Blackness and, 13, 72, 90; border policing and, 143; policing, 134, 150; underground of, 156; violence of, 77; whiteness and, 212n26. *See also* disidentification
identity, 13, 40, 138, 173–75, 204n16; gender, 181; national, 52; personal, 116; racial, 41, 52, 203n2, 216n6; searches, 78; whiteness and, 212n26
immigrants, 34, 45, 51, 53, 55, 95, 143, 214n12
immigration, 6, 41, 53–54; background, 34, 36, 48, 53, 68, 95; measures, 47; status, 51, 105

improvisation, 94, 152, 155, 185, 192, 219n36; black nonperformance as, 104; impossible, 183, 221n38; listening as, 93
inequality, 117, 209n83
interstitial listening, 8, 18, 24, 65, 91, 94, 121, 137, 145, 159, 167, 186, 190; borders and, 28, 67–68, 30, 162; transmutative, 98
invisibility, 41, 130, 206–7n4
Irigaray, Luce, 110–12, 115
iris scanning, 41, 161

jazz, 10, 205n29; festivals, 69
Joler, Vladan, 139–40
Jolo, Jérémie, 10, 59–64, 97, 99
jus sanguinis, 47, 107, 111
jus soli, 106–7, 209n55
justice, 3, 104, 133, 157, 201, 221n38; reproductive, 175

Kifferstein, Maria, 187–88
kinship, 46–49, 54, 57, 83, 85, 104, 106–7, 109–10, 197, 221n34; alternate practices of, 106, 215n20; authorized, 59; Black, 105–6, 113, 193; border policing and, 101; border zones and, 102–3; control of reproduction and, 116; population measurement and, 81; relations and, 119–20, 138; social register and, 66; sovereignty and, 141

labor, 14, 103, 111–12, 172; of mothering, 120; nonwhite, 52
Lacks, Henrietta, 114–15
language, 119, 138, 141, 155, 178, 192, 218n22; blackness and, 205n39; Black Swiss and, 3, 34–35; English, 169; French, 44; German, 82; music and, 164; national, 55; as proper knowledge, 23; regions of Switzerland, 5
Lavanchy, Anne, 47–48, 50
Leung, King-Ho, 176–77
Lewis, George E., 10, 189
life, 3, 5, 12–15, 21, 23–24, 40, 61, 67, 72, 76–80, 85, 87–90, 102–4, 107, 113–15, 118–22, 139–40, 149, 155, 157, 167, 171, 173, 188, 191, 195, 197–98, 221n34; authority over, 30, 71; blackness and, 41, 159–60, 169, 185, 194; commodification of, 111; human, 45, 56, 74, 82, 115, 147; ideals of, 97–98; listening and, 150; measurements of, 190; non-Black, 182; queer, 91–92; silence and, 8; social, 117; sovereign, 86; stolen, 152; Swiss way of, 55; technologies and, 161; working-class, 52. *See also* bare life; Black life; proper life
listening, 1–3, 5, 8, 12, 16–18, 22, 27, 64, 67–68, 138–42, 144, 150, 155, 157, 160–61, 190; archive and, 218n25; black noise and, 152; *Bodycam Exhibit 3* (Uzor) and, 20, 23–26, 28–29; borders and, 98, 163; doubling of, 158; fugitive, 139, 153, 218n3; as improvisation, 93; Jolo's work and, 61–62; modalities, 149; music and, 17, 165; ownership and, 150; racialization and, 218n23; refiguring of, 123, 138; silence and, 152–53; technology, 139–41, 160. *See also* interstitial listening
livity, 102, 197
Lorde, Audre, 120, 214n12
Lucerne Festival, 187–88, 198, 206n7, 222n2
lynching, 19, 25–27, 29. *See also* Floyd, George

Malcolm X, 16, 214n14
Malevich, Kazimir, 128–31, 217n24
Marriott, David, 177, 179, 210n96, 217n31, 217n34, 222n14. *See also* corpsing
Martinot, Steve, 25, 199
Mbembé, Achille, 84–87, 211n3, 223n9. *See also* necropolitics
Mennel, Kelechi Monika, 39–40
Michel, Noémi, 43–45
migrants. *See* immigrants
minstrelsy, 111, 182
miscegenation, 49, 112, 115
mixedness, 48–50
modernity, 2, 80, 86, 96, 140, 183, 189
Moten, Fred, 105–6, 118, 130–31, 134, 191–94, 204n24, 210n5, 210n11, 210n14, 221n45; affirmation and, 222n14; on blackness, 205n35, 205n39; nothingness and, 176–77; unasking, 210n14
motherness, 101–2, 109–10, 112, 116, 146
multiculturalism, 6, 44, 49–51, 219n25
musicians, 1, 8, 19–20, 30, 60–63, 122, 161, 163, 187, 190; Abrams and, 204n26; Black, 188–89; Black Swiss, 9; of Sun Ra Arkestra, 76
Muselmann, 81–82, 86

names, 7, 14–16, 105, 138, 149, 158, 169, 221n34; archive and, 168; borders and, 119–20; of identity, 175; kinship and, 141; pitch as, 160; unholdability of, 12; unthought and, 15

naming, 7, 12, 14–16, 90, 138, 147, 163, 169, 175; unnaming, 90, 119–20, 176

narrative, 21, 95, 97, 135, 147, 149, 157; of causality, 201; of emancipation, 148, 152, 158, 165; form, 29–30, 78, 86, 142–44, 154, 159, 165; history and, 22, 214n10; national, 42; structure, 142–43, 169; temporal, 45, 78

natality, 112–13, 116–18

natalpolitics, 115, 219n33

national belonging, 34, 59, 73–74, 119, 204n19; antiblackness and, 71; blackness and, 54; kinship and, 106, 116; proper speech as, 155; racism and, 43–44

nationalism, 10, 71

nationality, 45, 47–51, 76, 204n16

national origin, 48, 50, 204n19

nation-states, 107, 143; antiblackness and, 6, 96; blackness and, 41; slavery and, 102; sovereignty of, 77; women's reproduction and, 113

Nativ (Thierry Gnahoré), 68–69

Ndiaye, Noémi, 111, 114, 211n29

necropolitics, 84–86, 214n65

noncitizenship, 83, 105

nonperformance, 103–8, 119, 152, 179, 183–83, 221n34

normativity, 40, 56–57

nothingness, 76, 129, 134, 156–57, 172–74, 183–84; of appearance, 176; blackness and, 131, 172, 176–77, 183, 186; existence and, 222n15

notness, 135, 156, 173–74; of appearances, 134, 179, 184; being and, 171, 173, 185; of black, 76; of double consciousness, 169; of life, 89

Oguntoye, Katharina, 5, 16, 56, 95. *See also* *Showing Our Colors*

opacity, 60–61, 64, 177

Opitz, May, 5, 56. *See also* *Showing Our Colors*

ownership, 14, 64, 75; abstraction and, 163; antiblackness as right to, 172; of black people, 71; of borders, 90; of flesh, 146, 218n25; identification and, 176; knowledge and, 186; of life, 112; listening and, 150; of women, 113, 146

people of color, 18, 36, 138, 203n2

performance, 177–80, 182–83, 221n34; of antiblackness, 71, 163, 179; of blackness, 182; of desire, 177; deviant reproductive labor of Black women as, 113; of erotics, 111; Fanon's writing as, 212n20; identification and, 156; imitation and, 20; musical, 8; of race, 211n13; of whiteness, 50, 211n13. *See also* corpsing; nonperformance

Peter, Mike Ben, 3, 17–18, 22, 28, 30, 122, 166, 201

phonorealism, 28, 162

pitch, 8, 23, 60, 160–61, 163; mapping, 27

police, 3, 18–19, 24–25, 36–39, 101; in *Bodycam Exhibit 3* (Uzor), 20, 30, 163; borders and, 125; as exception, 199; identification and, 72, 152; profiling, 36, 46; sovereignty as, 165; violence, 199–200; white people and, 87. *See also* Chauvin, Derek

policing, 31, 42, 90, 97, 120; archive and, 142, 144; bias against nonwhites in, 37; blackness, 69; identification and, 134, 150; of motherness, 101–2. *See also* border policing

political, the, 57, 77, 80, 96, 116–19, 170, 212n29; limits of, 85; music and, 123; racializing process of, 86; speech and, 150

proper, the, 81, 90, 97, 125, 146, 148, 203n2; critique of, 139

proper belonging, 9–10, 54, 59–60, 72–73, 109, 193; borders and, 49; culture and, 97; enforcement of, 66–67; identification and, 36; listening and, 150; policing of motherness and, 102; women's wombs and, 146

proper life, 79–80, 83–84, 87, 90, 103–4, 144, 160, 179–80; Black lives and, 125; citizenship and, 89, 98; death and, 86, 88–89; fabricated, 2; immigrants and, 143; women and, 111–13, 115

property, 14, 25, 106, 111, 141, 149; knowledge as, 186; law, 105; space of, 146

purity, 49–50, 89, 133

queerness, 91, 181

race, 13, 15, 18, 25–26, 40, 65, 81–83, 86, 89, 151, 189, 207n25; Afro-Germanness and, 204n16; antiblackness and, 205n34; citizenship and, 119; emancipation and, 29; erasure of, 37–38, 46, 50, 181; France and, 52–53, 208n39; Italy and, 209n55; listening and, 219n2; mixedness and, 48–49; music world and, 123; natality and, 116; performance of, 211n13; proper life and, 79; sociology of, 152; surveillance technologies and, 161; Switzerland and, 6, 34, 36, 41–42, 44–46, 52–54, 57, 68, 174; tokenism and, 205n33

racelessness, 44, 49, 207n26

racial identity, 13, 34, 41, 52, 216n6

racialization, 6, 13, 37, 39–40, 42, 50, 57, 96; authenticity and, 86; biopolitics and, 82; border control and, 45; citizenship and, 54, 83; erasure of, 36, 148; homo sacer and, 80; listening and, 150, 218n23; reproduction and, 116; Swiss nationality and, 48

racial profiling, 3, 18, 35, 39, 41, 46, 98

racism, 13, 15, 26, 36–39, 56, 82–85, 114, 207n25, 214n15; antiblack, 3–4, 6, 18, 36, 38–39, 81; anti-racism, 25; Black women and, 175; naturalization and, 207n24; in Switzerland, 38–39, 41–46, 49–50, 52–55, 68, 98, 174, 180–81, 204n19, 214n12

realness, 26–27, 73, 160

recording technologies, 27, 161

refugees, 34, 98, 105, 193

Reinhardt, Ad, 130–34

relationality, 64, 106

reproduction, 47, 109, 111–16; of race, 219n2; sound, 28

resources, 139–41, 143–44, 146, 157, 181, 191, 218n25

right to kill, 77, 79, 84–85, 87–88

Schultz, Dagmar, 5, 56, 95. *See also Showing Our Colors*

Schwarz-Sein, 169–73, 177–78, 184–86, 190–91, 194

segregation, 51, 118

sexism, 175, 180–81

Sexton, Jared, 25, 176, 199, 209n65, 214n65, 222n14, 223n9

Showing Our Colors (Oguntoye, Opitz, and Schultz), 5, 56, 83, 95, 102

silence, 2, 8, 52, 119, 138, 143, 152–54; in *8'46" George Floyd in Memoriam* (Uzor), 122–27, 132, 173–74; in *Body-cam Exhibit 3* (Uzor), 24–26; protest of, 217n16; segregation and, 206–7n4; of the unknown, 156; unnamed and, 165

singularity, 14, 90, 154, 156, 159, 166–69

slavery, 55, 102, 108, 148, 189; abolition of, 158; antiblack, 29; *Betty's Case* and, 103–5, 107; Blackness and, 138; Black women's reproduction during, 112; political power and, 151; property relations and, 14; Switzerland and, 53

slaves, 53, 102, 106–7, 147–48, 156; emancipation of, 189; owning, 104, 111–12, 146, 151; progeny of, 57; suicide of, 87; value of, 9

slave trade, 53, 204n23

sound recording, 20, 23, 26–27, 91, 161–62

sovereignty, 56, 77–78, 84–86, 88, 104, 143, 146–47, 165–66, 190, 194, 212n29; citizenship as, 80, 85, 89, 145; national, 54, 113; nationalism and, 71

speech, 21, 116, 139–41, 145, 155, 157, 162, 167; archive and, 168; imitation of, 161; listening technologies and, 139; recognition of, 28; subaltern and, 150–51

Spillers, Hortense, 105, 109, 146–48, 215n20

Spivak, Gayatri Chakravorty, 150–51, 191

Stadler, Gustavus, 26–27

state of exception, 77, 79, 81, 85, 88, 147, 212n32

Stovall, Tyler, 52–53, 108

subjectivity, 55, 129, 170; Black, 83; modern, 111

Sun Ra, 14, 65

Sun Ra Arkestra, 76, 173

Swissness, 47, 50, 54, 57, 90, 97, 140, 190; antiblackness and, 3; Black, 7, 34–35, 40, 68, 192; Black lives and, 36, 189; Blackness and, 55, 188; displacement and, 193; motto of, 168; music and, 10; political performativity of, 45; refusal and, 192

Swiss People Party (SVP), 42–45

Switzerland, 2–4, 6, 9, 33–34, 36–37, 43, 45–47, 49, 51–53, 55–57, 96, 184; bells and, 189; Biel/Bienne, 60; Blackness in, 6, 68–69, 81, 95, 98, 123, 140, 144, 155, 188; Black women in, 94–95; borders of, 36, 68, 80; citizenship and, 104–6; feminist organizations in, 214n12; Geneva, 43–45, 91; Gymnasium in, 19; musical pedagogical system in, 205n29; National Council, 42, 182*f*; racism in, 13; St. Gallen, 11, 18; Zurich, 36, 39, 135. *See also* antiblackness: in Switzerland; Black lives in Switzerland; racism: in Switzerland

Taylor, Cecil, 130–34
technology: AI, 139–40, 144; black as ancient, 137; borders as, 109; iris scan, 161; listening, 139–41, 160; modern, 41; sound recording, 27; space travel, 91
Thirteenth Amendment, 107, 116
togetherness, 91, 94, 139, 221n38; Blackness and, 15, 56–57, 61, 63, 99, 182, 195; of human affairs, 117; as nothingness, 157

unthought, the, 9, 15, 64, 138, 174, 181, 184, 188
Usländerdütsch, 15, 155
Uzor, Charles, 11, 18–19, 23, 25–28, 125, 167, 174, 188, 199, 201, 209n56; *8′46″ George Floyd in Memoriam*, 19, 122–24, 126–27, 131, 135, 173–74, 198; *Bodycam Exhibit 3: George Floyd in Memoriam*, 18–24, 28–30, 101, 159–63, 166–68, 184, 198

violence, 4–5, 27, 87, 148; antiblack, 14, 26, 101, 122, 198–201, 213n53; of border policing, 125, 213n53; of identification, 77, 190; naming and, 7; police, 199–200; sexual, 114; structural, 78; against women, 112, 116

Wa Baile, Mohamed, 3, 36–39, 90
Weber, Samuel, 80, 88–89
Weheliye, Alexander, 35, 81–82, 86, 147, 169–71
Wheatly, Phillis, 108–9
white gaze, 11, 57, 61, 72–78, 89, 98, 125–27, 131, 134–35, 191, 212n22; of the archive, 163; blackness and, 104, 177, 185; doubling of, 179, 184, 192; failure and, 177, 179, 183
whiteness, 25, 45, 73, 82, 87, 132, 134, 209n53; antiblackness and, 125; biometrics and, 161; blackness and, 13, 54, 72, 114, 169, 171; classical music and, 205n29; identity and, 212n26; multiculturalism and, 44, 50; as norm, 205n33; performance of, 50, 53, 211n13; reproductive capacity of, 112
white noise, 124–27, 133, 152
white supremacy, 3, 25
Wien Modern, 19, 206n25
Wilderson III, Frank B., 29–30, 87, 89, 125, 210n14, 213n53
women, 110–16, 146, 158, 181; Aboriginal, 113; Afro-German, 102, 180; Afro-women, 16; American, 109; Black lives and, 95; of color, 180; Puerto Rican, 175; reproductive rights of, 113–16, 120, 216n35; violence against, 112, 116; white, 56, 111, 180; women's movement, 214n12; women's rights, 96. *See also* Black women
Women of Black Heritage (WBH), 39, 203n5

Ziarek, Ewa Plonowska, 113–14, 116